RUSTY, A NOBLE RACCOON. WE LEARNED MUCH FROM ONE ANOTHER—
SOME OF IT THE HARD WAY.

Our Small Native Animals

Their Habits and Care

by ROBERT SNEDIGAR

Curator of Reptiles, Amphibians and Invertebrates

Chicago Zoological Park

Brookfield, Illinois

REVISED AND ENLARGED EDITION

DOVER PUBLICATIONS, INC.

New York

Published in Canada by General Publishing Company, Ltd., 30 Lesmill Road, Don Mills, Toronto, Ontario.

Published in the United Kingdom by Constable and Company, Ltd., 10 Orange Street, London W.C.2.

This new Dover edition, first published in 1963, is a revised and enlarged edition of the work first published by Random House in 1939.

Standard Book Number: 486-21022-7
Library of Congress Catalog Card Number: 63-19501

Manufactured in the United States of America
Dover Publications, Inc.
180 Varick Street
New York, N.Y. 10014

Preface to the Dover Edition

WHILE THE text is basically the same as in the first edition of 1939, several new features add to the usefulness and appearance of this new 1963 edition. Chief among these new assets are the dozens of new clear and authentic photographic illustrations. A new chapter, "Twenty-three Years After," touches on some of the developments of the last twenty years in our knowledge of animal behavior. New material has been added on the coyote, on animal ailments, and on the dangers to wildlife from insecticide sprays. All scientific names have been checked and brought up to date where necessary. Other minor corrections have also been made to make everything in the book conform to 1962 knowledge and practices. A brief bibliography has been included.

I sincerely hope that in its bright new format this book will continue to bring guidance and enjoyment to a growing number of animal lovers and pet owners.

Through the years that have slipped past since the original edition of this book, I have been fortunate in that the entire time has been spent in working with live animals. To those animals, I am greatly indebted not only for amusement and addition to my knowledge of their ways, but occasional salutary bits of discipline when I offended.

In addition to these obligations and the general ones I owe my co-workers in American zoos, I am, in the preparation of this revision, particularly indebted to Ronald L. Blakely, George B. Rabb and Weaver Williamson of the Chicago Zoological Park for aid in preparing new material and checking old in their respective fields of ornithology, animal behavior and ecology and veterinary medicine.

New photographs are from many sources. D. Dwight Davis, Lewis W. Walker, the National Park Service and the Fish and Wildlife Service are prominent among them. Naturally, I am greatly obliged to the Chicago Zoological Society and our director, Robert Bean, for the freedom I have been permitted and the splendid opportunities the Park has afforded me to observe and learn about animals.

ROBERT SNEDIGAR

Brookfield, Illinois
January, 1963

Foreword

THE STUDY of Natural History has come to be a recognized activity in camps, parks and schools. The Boy and Girl Scouts, the Camp Fire Girls and other similar organizations, the National Parks, many State and City Parks, thousands of summer camps, as well as public and private schools, have well-established programs of nature study. In this great and fascinating field, including, as it does, everything that is or happens according to natural law, live animals take first place in the interests of the majority of amateur naturalists, especially young people. In the schools of our large cities where it is necessary to "bring the country to the city child," the problem of procuring and keeping live animals is a serious one. However, thanks to interested individuals and organizations, it is being solved in at least some of the schools.

With this intense and sincere interest in the behavior and habits of live animals, comes the need for first-hand, authoritative information about what wild animals can be kept and how they should be cared for in the school nature room, the camp museum, or at home. Proper housing and care are obviously essential; and to that end it is necessary to have accurate knowledge of these pets.

In *Our Small Native Animals: Their Habits and Care*, Mr. Snedigar presents a very useful book. His thorough knowledge of these creatures was acquired first of all out of doors, where the real naturalist spends most of his time. Further observations were made on animals reared under the most favorable circumstances in environments approaching as closely as possible their natural habits. It is refreshing to find a treatment of the subject that is entirely sympathetic without being in any way sentimental. The accounts of life histories and habits of the animals are never dull and are frequently very entertaining. The information about housing and feeding is explicit and complete, the result of years of experience in caring for many different small native animals.

ADA KNEALE BURNS
Director of Staff, School Nature League

Acknowledgments

IN THE MAKING of this book, I have found it necessary to draw freely upon the experiences of others to augment my own store of knowledge. My debts are too numerous to itemize fully, but certain obligations are so large as to demand recognition here.

My chief, Dr. G. K. Noble, generously gave permission for the use of material and formulas developed in the Departments of Experimental Biology and Herpetology of the American Museum of Natural History.

C. M. Bogert and Clifford H. Pope advised in the preparation of the sections on reptiles and amphibians and checked these in proof. The experience of several summers in charge of the Lake Kanawaukee Regional Museum of Harriman State Park made the assistance of John C. Orth most valuable in determining the scope and contents of the work and in its preparation. W. G. Hassler gave general advice and help and, in addition, did most of the collecting and preparing of the photographs. G. H. H. Tate kindly read the proof of the section on mammals. Miss Margaret McKenny and John T. Nichols made many useful suggestions with regard to the birds and checked that section in proof. Mrs. Ada Kneale Burns read the entire manuscript and suggested ways in which it could be made of greater usefulness to the nature student and teacher.

Copeia, the organ of the Society of Ichthyologists and Herpetologists, and the *Journal of Mammalogy* provided useful items and ideas. The illustrations of bird houses and feeding trays were drawn from models in the offices of the Audubon Society and this organization's bulletins were freely consulted and used in the preparation of the bird section. The text drawings are by Alma W. Froderstrom.

To these and the many unnamed others who have given of their help and knowledge, I can only return my thanks and hope they will find recompense for their labors in the furtherance of a better understanding of our small animal friends.

R. S.

Contents

xi

Photographic Illustrations

Illustrations in Text

Introduction

THE ANIMALS most acceptable to man have always been those whose behavior he could interpret in terms of human psychology and whose virtues and intelligence most closely approached human traits. The dog, with faithfulness, trust and unswerving loyalty, has, by the persistence and humility of his affection, made for himself the place of first companion to man. Our appraisal and valuation of the other creatures has always been based upon their assumed possession of human virtues, intelligence and feelings. The lordliness of the lion, the wisdom of the owl, the stupidity of the donkey, the aristocratic disdain of the cat, the slyness of the weasel, the industry of the beaver—what are these? Simply descriptive terms based upon an unwarranted assumption that the mental habits and reactions of animals are like our own. Actually, their thinking and the resulting action is of a different sort from ours and not in the least subject to the same judgments. Naturally, animals like the dog, understandable in our terms and willing more or less to accept our standards of behavior in place of their own ancient and primitive law, are our favorites. Other creatures, refusing to give up their habits and ancestral traits in captivity, and carrying on their age-old patterns of life in defiant independence of our human feelings and regulations, we call wild, vicious and untamable. Instead of liking and appreciating their integrity of character, we are disappointed that we cannot warp them into caricatures of human beings.

If animals—and many of them do—respond to human gentleness with toleration and confidence, well and good. If they have no liking for human company and refuse to be petted or to sit up and beg, we may as well try to appreciate their point of view and be content to watch them and to try to understand the factors underlying their unsocial behavior.

Most animals in captivity are not insistent upon their rights under the old law of tooth and claw and, in so far as their human associates are concerned, waive them. But they seldom see any valid reason why their enemies and their prey of gen-

erations should be included in the truce. This leads to misunder-
standings and difficulties. The owner of an animal is sometimes
disappointed when he finds that his "pet" is lacking in certain
virtues the animal should not be expected to possess. Perhaps a
grasshopper mouse, gentle and unafraid of his owner, full of
pleasing mannerisms and tricks, falls from grace by slaughtering
and devouring the white-footed mouse put in his cage as a companion. The tiny butcher's lack of a conviction of sin and his
willingness to hop upon his chagrined master's hand and, sitting
there, clean the remains of bloody evidence from whiskers and
feet, adds an extra touch of callous guilt. But—grasshopper mice
have for uncounted generations been preying in this same bloody
fashion upon this very game. The culprit's attitude in the matter
is likely one of thankfulness—felt in a comfortable distention
of stomach rather than expressed—at his owner's consideration
in giving him such an excellent dinner; just the sort of dinner he
and his family most relish.

There are animals which fiercely and ceaselessly resent captivity and its restrictions, but these are in the minority. Under
proper captive conditions most wildlings enjoy a life uncom-
plicated by fear and relentless hunger. Visitors to the zoo are
often guilty of slightly misplaced sympathy for the inmates. In
their eyes those lords of the jungle, the lion and the tiger, storm
behind their bars and roar in rage for their stolen freedom.
Do they? Or do these lords—in much better condition and coat
than large cats in the wild and much more likely to attain a
generous span of years—squall petulantly and pace the cage
only because the keeper is due with the daily horsemeat?

The business of live animal-keeping is a sort of tightrope
walking, a narrow path between evils. On the one hand lies the
sticky slough of sentimentality; on the other the thorny and
barren ground of cruelty and callousness. Between the two is a
small and fruitful area; the narrow ground of respect and a
realization that our animals have the right to as nearly normal
a life as the restrictions of captivity can permit.

A normal life means a healthy life. And with that up crops
the great bugaboo of the animal keeper: how can animals be
kept healthy?

Words easily dispose of it. *An approximation of the natural
foods and environment is often essential, always desirable, in the
keeping of wild animals in captivity.* A nice formula, easily said
and, if carried out, one that works. But, like all good formulas
that work, its carrying out takes a sight of doing, and much

care and ingenuity are needed for even its partial fulfillment.

It is easily seen that a due consideration and application of it will limit the amateur's collection to those creatures for which he can provide the proper living conditions and foods. Such limitation will, in the long run, save expense and trouble and much needless animal suffering. Monkeys, other small tropical mammals and exotic birds make wonderfully attractive pets, but the requirements for their successful maintenance are so special and their needs so peculiar that they are hardly advisable for the home, school or camp animal family.

These collections are better made up from the great body of our own native species. They require no adjustment to a strange climate, are easier to feed and rear and, once their acquaintance has been made, are just as attractive as the exotics. They are not at all difficult to get. The common experience has been that as soon as word gets around that animals are wanted and will be cared for, a procession of babies and cripples begins to pour into the new establishment. The cripples are difficult—often death is the only kind solution—but young animals raised by hand tame readily and become the most charming of pets.

Such a collection of native animals, if at all extensive, fits into the scheme of present day natural history education by showing a cross section of the wild life of the region.

This educational value is heightened if, where possible, each species is given its rightful plant associates and natural physical surroundings. Where such provision has been made for the animals are more likely to be resented than those familiar with possible and really fine photographs may be made. These, in addition to being of pleasure and interest to the immediate observer and his friends, may have a genuine scientific value. There are many gaps in our knowledge of the habits and character of some of our small creatures which can only be filled by the intelligent notes of interested amateurs.

Along with the great problem of the welfare of the animals, there is another that demands the immediate attention of the amateur zoo keeper: *people*. Unless a great deal of tact and care are exercised, things are very likely to happen when people and animals are mixed too freely. For the zoo owner, his friends and neighbors, a few rules strictly adhered to are advisable in order that these little happenings may be avoided. For one thing, gently but firmly refuse to permit casual visitors to play with or handle your animal guests. Persons unaccustomed to handling animals are more likely to be resented than those familiar with their "feel," and you are legally responsible for any injury a visitor may receive. In this connection, it may be said that human constitutions vary and what to one person is an insignificant nip may to another, through infection or shock, become a serious matter. Needless to say, any animal bite or scratch should be considered as infected and properly treated at once. Keep iodine handy and use immediately upon any injury, no matter how seemingly unimportant.

For the peace of mind of the family and neighbors as well as for his own mental comfort, the zoo owner should make every effort to keep his pets caged or penned securely. There is little ground for excuse in case of an escape. Even the most harmless creatures have scare possibilities. An undeodorized skunk may be, and often is, in his own home a quite gentle, harmless and affectionate pet. But strolling through your neighbor's living room, he is apt to be startling and, if himself startled, disconcerting and likely to put an unpleasant complexion—and odor—upon the situation.

It is not difficult to become acquainted with animals. Nor, contrary to popular opinion, are their scientific names beyond the grasp of intelligent young people. Popular names, because of their indefiniteness and the fact that often several quite different creatures may be known by the same term in different localities, are liable to cause confusion. A case in point is the word "scorpion." Properly applied, it refers to a certain group of stinging creatures of the spider family; but in the southern states, it is used in referring to a group of lizards.

In an attempt to pin definite names which closely indicate actual relationships upon the animal and plant life of the world, scientists of all nations use the system of nomenclature originated by the Swedish botanist Linnaeus in the eighteenth century. It has become necessary to make some changes in the use of the *Systema Naturae* of Linnaeus, but the principles of naming and describing animals which he laid down are one of our basic scientific tools today. By using them we may designate *exactly* the form we mean without danger of confusion.

For an analysis of a scientific name, let us take that of one of our red squirrels. Popularly, this animal is known variously as the chickaree, the bummer squirrel, the mountain boomer and the red squirrel. (Related forms, similar but with differences obvious even to the unpracticed eye, also go under these same names.) Scientifically, this squirrel is *Tamiasciurus hudsonicus minnesota*, and the name is applicable to no other form. Scientific names are usually made up of Latin or Greek descriptive terms. The first word is the name of the genus or group the animal belongs to. The second term, usually an adjective, is the name of the particular species. In some cases species are further divided into subspecies. These latter are ordinarily regional forms and the name applied ideally is that of the particular locality. This, however, is not adhered to and we find subspecies with names derived from several sources. In the case of the red squirrel, *Tamiasciurus hudsonicus minnesota,* the generic term, *Tamiasciurus,* is taken from the Greek and means "shade-tail steward," an obvious and apt name for this group of animals. *Hudsonicus,* the specific name, was given because the species was first described from the Hudson Bay Region. The subspecific term, *minnesota,* is obviously a locality reference.

What it all amounts to is this: generic differences are differences in fundamental structure. Specific differences—sometimes very obvious and sometimes hard to see—are chiefly differences in external characters. Subspecific differences are ordinarily those of color, size and range.

The various states differ in the laws protecting wild life. The possession of a certain animal in one state may be perfectly all right but, across a boundary line, be illegal. This applies principally to birds and mammals, but in certain cases to reptiles and amphibians as well. For example, the box turtle and the wood turtle are protected in New York State and it is illegal to have them in your possession without a permit.

Rodents are almost universally regarded as pests and there are no restrictions on the keeping of the smaller ones. The same is true of snakes, many of which are of considerable value in the control of these same rodent pests, but are not protected by law. If, however, snakes escape—this has happened—through carelessness and terrorize a crowded neighborhood, the keeper may find himself involved with the law in a quite different fashion. Under charges of "having allowed a dangerous animal to escape," he will be forced to appear before a judge and explain.

In the matter of birds, squirrels and the larger mammals, the keeper, if he wishes to avoid trouble and perhaps the loss of an animal to which he has become attached, will do well to familiarize himself with the regulations of his state and the necessary steps for getting the proper permits to keep protected material. The Conservation Department of your state will be able to give full information.

Squirrels and Chipmunks

FROM THE amateur animal keeper's point of view, each type of wild life has its own peculiar set of virtues and paralleling vices. The mammals have, generally, the virtues of good looks, attractive behavior and a capability for engaging and returning the affection of man. The reverse of the picture is that they are more exacting in their need for routine attention, feeding, cleanliness and warmth, than some of the more easily tended, less intelligent lower animals.

Most of us have no facilities for the housing and care of large livestock and would be at a loss to provide quarters for anything which meets the popular conception of a "wild animal." However, as one looks through the wrong end of a telescope and sees the world marvelously dwarfed, so we, in watching the behavior of a group of small creatures, may see in miniature all the traits of the animal kingdom.

These small creatures need not be strange to be interesting. Indeed, some of the most common animals have the largest and most amusing bags of tricks for our enjoyment and education.

Of all the animals of North America, the squirrels are perhaps the best known and most often seen in a state of nature. The varied climatic conditions of our continent have made possible the development of a number of principal species and many geographic subspecies, each of the latter peculiar to its own region.

THE RED SQUIRREL

With the largest range of all, the red squirrel, *Tamiasciurus hudsonicus*, or one of its close relatives, is found throughout practically all of our forested areas. Extremely adaptable, the red squirrel leads a comfortable life in the mountains of Alaska as well as in the southern Alleghenies. Unlike many other mammals of cold climates, he does not pass the winter idly in hibernation. Snow and cold weather are no bar to his activity and no discouragers of his curiosity and impudence. The ski novice, sliding down an Adirondack slope, making jumps that turn into awkward tumbles which he hopes have not been seen, is very likely to find his lack of skill the subject of sarcastic and scolding comment from a couple of red squirrels flirting around the trunk of a snow-heavy tree.

The Ojibway Indians called the red squirrel Adjidaumo, "tail-in-air." The name is apt. His tail is always in the air and seldom still. Adjidaumo finds his tail invaluable in the expression of emotion and opinion. He jerks it, twitches it, and waves it, and by its flirtings and flauntings adequately conveys a good idea of his state of mind at the moment. On his home grounds, he never seems to have a very cordial attitude toward an intruder. He chatters and barks, sputters, scolds, whistles and shrieks; ranges vocally through everything from a mild excitement to a wild frenzy.

But undisturbed, and with no excuse for scolding, the red squirrel almost sings. From his call, a series of rolling vibrant notes, *tcher-r-r-r*, ordinarily diminishing in pitch and speed of delivery, has come the popular name of chickaree.

The red squirrel prefers to stay off the ground and, except for food storing and foraging, making a quick trip from one tree to another or flying in frantic and screaming fear from one of his fellows, stays aloft. Often the nests, especially those used in summer, are built among high branches. Twigs, leaves, shredded bark and other forest litter are the materials used for the rough outer construction. Inside, soft dry grasses, dry moss and leaves make it habitable and warm. For winter nests, the cavities of hollow trees and stumps, or cracks and crannies in the rock, are favored locations.

The young are born in late spring or early summer and by the latter part of August are seen about the family domain, helping and hindering in the task of getting winter food into storage. Apparently, the young often leave the parents as soon as they know their way about, but one observer who had the good fortune to keep the same group of chickarees under his eye for several seasons, reported differently. Mr. M. A. Walton, in *A Hermit's Wild Friends*, said:

"Young squirrels remain with their parents the first winter, but in April the female turns the family over to the male and makes another nest of moss, leaves and dry grass in the top of a tall pine or hemlock tree. While she is engaged in new duties the male looks after the young squirrels that are now full-grown.

He finishes their education and locates the young males on territory which they ever after hold."

Their holding of these territories is not unaccompanied by discord and strife. Little birdies in the nest agree—sometimes— and perhaps little squirrels also. But grown-up male chicka- rees violently, noisily and, too often bloodily, resent visitors of their own kind within their selected areas. As with many other animals possessed of an exaggerated sense of territorial rights, a chickaree out of his own bounds seems afflicted with an in- feriority complex that makes him fly, shrieking and fear-pan- icked, from a smaller squirrel that he could easily whip on neutral ground.

In nature, the red squirrel eats almost anything available. After a winter of dependence upon stored foods—nuts, seeds of pine, spruce and other coniferous trees, dried mushrooms and so on—the berries of wintergreen and the partridge berry re- vealed by the melting snow are eagerly devoured. The tender buds of trees at this time offer an appetizing change of diet. The chickaree has a sweet tooth and spring permits him to indulge it. As the sap rises in the sugar maples, the squirrel seeks out the ends of wind-broken branches and laps up the bleeding sap. If he doesn't find such natural sources, he may score the bark and lick up the sweets from the wound. As the season progresses, other foods become available. Insects, stray bits of meat, even bird eggs and young, leaves, flowers and fruits set the chickaree table. Mushrooms are a favored item, and discarded pieces of them litter the stumps of his haunts. Many of them are har- vested, cleaned, dried and stored for future use. Although a number of species of mushrooms are deadly poison to human beings, and certain others have very bad, although not fatal, effects, squirrels are able to eat them and seem to have complete immunity to their poisons.

The large and deadly orange-capped fly mushroom, *Amanita muscaria*, is an especial squirrel favorite. Because of its likeness to a delicately flavored and rare edible mushroom, Caesar's mushroom, *Amanita caesaria*, the fly mushroom is often mis- taken for it by man with fatal results. The liking of the squirrels for the fly mushroom has added to the confusion and danger. Several seasons ago, an acquaintance insisted to me that he knew of a place abundant in the rare *caesaria*. Fortunately, he was not fond of mushrooms and still less fond of his own judgment, for his abundance turned out to be of the poisonous *muscaria*. His defense and the basis of his mis-identification was that squirrels were eating the mushrooms and so he had assumed them fit for man.

Late summer finds the chickaree gathering these and many other fungi, together with all sorts of seeds, especially those of the coniferous trees, and caching them about his domain. The chickaree puts his trust in no one basket. His immediate relatives are ready at any favorable time to raid his storehouse. His larger cousin, the gray squirrel, is ready to brave the chickaree's wrath for the sake of a few nuts, while that thief of the woods, the blue jay, even more impudent and saucy than Adjidaumo himself, welcomes any chance to make a crafty raid. Man often diverts the chickaree's providence to his own profit and there are recorded cases of as much as a bushel and a half of nuts being taken from one red squirrel store.

The *hudsonicus* group, of which the eastern chickaree is the principal representative, is the most extensive in range, and differs from the two western species, *douglasii* and *fremonti*, and their many subspecies, mostly in color. In habit and behavior, the chickarees are much alike and in spite of color differences their relationship is obvious. The Douglas chickaree of the Northwest is particularly noted for its musical call, a call so melodious that it is commonly mistaken for that of a bird. Otherwise, he is just as noisy as his eastern relatives and equally industrious in gathering and storing quantities of winter food. First among these stores are the seeds of the coniferous trees of the great forests in which he lives. A considerable degree of intelligence is displayed in their storage. The cones are cut while still green and at once cached in wet, soggy ground. Cones, given a chance to dry, disintegrate and the seeds are lost. Kept damp in this fashion, they remain intact until the chickaree needs them.

In captivity, red squirrels are difficult. Possessed of considerable individuality, their behavior has been described in widely varying terms: timid and shy; aggressive and noisy; easily tamed; impossible to tame; vicious, gentle, dumb, intelligent; these small creatures fairly stagger under a burden of behavior adjectives. Naturally, their age at time of capture has a great deal to do with later character. Adults adjust to the restrictions of cage life and human companionship with difficulty. Youngsters, half-grown, or taken from the nest and weaned on a bottle, adapt themselves more easily and often come to regard their owner with friendliness instead of the traditional and scolding distrust.

Courtesy American Museum of Natural History

DOUGLAS RED SQUIRREL

U.S. Fish and Wildlife Service, photo by E. P. Haddon

GRAY SQUIRREL

ABERT TUFTED-EAR SQUIRREL

KAIBAB TUFTED-EAR SQUIRREL

If the adjustment to captive life is successfully made, the owner may find that it has been almost too successful and that the chickarees have taken over the premises. And—once in possession, they tolerate very little nonsense. A pair that I kept —I should say, that permitted *me* to live in the apartment with *them*—while adding considerably to the uncertainty of life, made me one of the family in a somewhat unflattering fashion. They had no regard for human rights and fought one another for the privilege of screaming at me. The moment the door opened, they began to scold and demand to be let out. Released, they raced up and down the curtains, under the couch, over the bookcases and desk, in and under everything. In particular, they delighted to scramble over any human being unfortunate enough to get in the way. Ordinarily, I didn't mind. When they raced madly around a terror-stricken friend's too ample waistline and finally took sulky refuge in his pocket I thought the situation—and my pets—very amusing. My own scratches seemed a small sacrifice. Unfortunately for their continued freedom and my good feeling toward their little tricks, I came to their notice one evening as I was preparing to retire. The game of tag was announced by two flying leaps for me. The only thing I could do was to lie down on the floor. Both of them sat on my chest, lecturing me because I wouldn't stand up like a tree and play, the while I scolded them in a small and bitter voice about their manners and wondered about the total footage of my scratches.

In spite of their familiarity and impudent behavior, this pair could not be described as tame. While unafraid of people, they resented handling and, if held too closely, used tooth and claw.

It goes without saying that creatures as highly nervous in temperament as the chickarees must be treated gently and shielded from unusual excitements. Even exceptionally docile individuals, taken young and carefully reared, cannot be expected to become thoroughly tame and safe. Perhaps one person only will be admitted to complete friendship and others still remain objects strange and fearful.

THE GRAY SQUIRREL

Size has little to do with scrappiness. The gray squirrel of the East, *Sciurus carolinensis,* shares a part of the red squirrel range and, being twice the size of the chickaree, might well be expected to be squirrel boss within the shared regions. On the contrary, the chickaree, with savage and frenzied onslaught, puts to rout any gray venturing within its chosen territory. Occasionally, a gray tries to stand and fight, but his only safety against the speed and biting attack of his small opponent is in flight.

Although a highly nervous animal, and one to be treated gently and protected from excitement in captivity, the gray squirrel is much less highly strung than the chickaree. This less hysterical disposition has perhaps had a great deal to do with the gray's consenting to share its territory with man, his automobiles, his band concerts and his picnic parties; for the gray squirrel is the squirrel of the city park and is familiar even to the most unwilling of naturalists. Any day clement enough for people to be out, find him bounding along gracefully, going from person to person begging shamelessly for the squirrel equivalent of the price of a cup of coffee. These half-tame squirrels lead a difficult life. In summer food is plentiful. Peanut vendors are all about and the walks are filled with strollers eager for entertainment. Picnic parties scatter remnants. Beneath the trees and in the borders are delicate, new-growing things and summer insects and their larvae. The squirrel is industrious and stores a part of this provender, the nuts particularly. A favored method of storage is to bury each nut in a hole a couple of inches deep and cover it over carefully. It is almost as if the squirrel were trying to plant a new forest. Indeed, in the wild where winter food is plentiful, many of these seeds are overlooked and left to sprout and grow. The squirrels have always done their share toward the perpetuation of the trees upon which their food supply depends. In city parks where winter food is scarce, stored food is not likely to be overlooked and the squirrel has a sharp need for every hoarded fragment. His scent is remarkably keen and he is able to hunt it out even through several inches of snow.

It is not an uncommon sight in the cold of winter mornings to see him sniffing hungrily along the frozen crust in search of his buried provisions. During these times, desperate and hungry, his attempts to wheedle stray humans into charity are pitiful. Since the park offers no natural food supply and squirrels are an interesting and lively accent in either summer or winter landscape, a steady and ample food supply should be given in exchange for their decorative service.

In nature the gray squirrel is largely arboreal and performs astounding feats of gymnastics among the higher branches. The home nest is, by preference, the hollow of a rotted tree trunk, but the squirrels often use a nest built of twigs, leaves, shredded bark and grass in the crotch of limbs or in the branches.

Less vocal than the chickaree, the gray squirrel commonly gives vent to his feelings in a husky bark. He is curious and likes to see what is going on, but because of his less excitable nature his curiosity is not accompanied by a running fire of scolding comment.

The gray squirrel accepts the conditions of captivity with more grace than his red cousins. This, of course, is a matter of individual temperament, but generally, grays readily become accustomed to the company of human beings and learn to depend upon them for the needs of life. However, they usually retain in captivity a certain vigorous independence and an inclination to do as they please. Often but one person is recognized as an intimate and permitted the liberty of freely handling and petting the captive. Jerry, belonging to John W. Thomson, Jr., is of this disposition. Although a winter resident of the School Nature League Headquarters, and a summer exhibit at the Kanawaukee Regional Museum in Harriman State Park for years, Jerry, in spite of a wide acquaintanceship, has never been genuinely friendly with anyone except his owner. This has been especially true of late years—Jerry is getting to be an old gentleman as squirrels go—and people too insistent upon his attention are very apt to get it—right in the finger.

Jerry has not always been captive. Three springs ago, he became seriously afflicted with a mange-like disease. His hair fell out; his fine, bushy tail became spottily naked and scabby; his eyes were inflamed; his temper bad and his voice querulous and complaining. Opinions and diagnoses of his disease differed. His years made plausible the idea that Jerry was suffering the natural results of old age. In a sad condition, he was taken as usual to Kanawaukee in the late spring. There, with the aid of John C. Orth, the curator in charge, intensive treatments aimed at the possibility of a dietary deficiency or an infestation of parasites were begun. But nothing seemed to help either Jerry's physical condition or his disposition. It was decided as a last resort to let him go out into the woods and seek his own salvation. He was turned loose one morning and for a number of weeks thereafter, nothing was heard or seen of him.

Then, after it had been definitely decided that he had succumbed to his disease or the onslaught of a predatory hawk or owl, a message came into Kanawaukee asking the boys to come to a camp a couple of miles down the lake and see what they could do about a plaguy squirrel that had taken the camp over.

It was Jerry, in the beginnings of a new coat, fat and fresher than ever and in no mood to take nonsense from Girl Scouts. Through some woodland secret of his own, he had found a cure and was ready again to take up life with the human tribe. He recognized his owner and, after a preliminary nip to assure John that it was no dream, contentedly went back in the car to Kanawaukee and his cage.

THE WESTERN GRAY SQUIRREL

The gray squirrel of the East is a handsome creature but is not as beautiful and striking in appearance as his western cousin, the California gray squirrel, *Sciurus griseus.* The latter is a larger animal of a lighter, clearer gray. From its color, and to distinguish it from the ground squirrels of similar size of the same regions, it is popularly called the silver-gray. One subspecies, *anthonyi,* found in the more southerly parts of the Coast Range, is curious because of its coal black feet. The home of the silver grays is in the forests of pine and the oak groves covering the western slopes. The nesting habits are like those of the eastern gray, but the California squirrel is less tolerant of people and more wary and shy. Within its range its numbers are in no place excessive and many localities, seemingly ideal for their residence, lack this entertaining presence.

THE TUFTED-EAR SQUIRREL

Another group of extremely handsome and large squirrels peculiar to certain regions of the West is that of the tufted-ear species, *Sciurus aberti* and *Sciurus kaibabensis.* These are without doubt the most showy of all our squirrels and it is unfortunate that they are restricted to the comparatively inaccessible and mountainous regions of the Rockies. They may easily be recognized by the bobcat-like tufts of long hair on the ears, their reddish backs and partly or wholly white tails. *Kaibabensis* is limited to the plateau on the north side of the Colorado River from which it takes its name, and is interesting as an example of a species which has attained a highly specialized coloring through long isolation and inbreeding.

Resident in pine woods, the seeds and branches of these trees form their principal food; but they also consume such typically squirrel provender as acorns, roots, mushrooms, and in season, the eggs and young of birds.

Although less deserving than the chickaree of the name of

common scold, the tufted-ear squirrels sometimes vent their feelings on the subject of intrusion in much the same fashion. Ordinarily, a chuckling or barking call is the indication of their presence.

THE FOX SQUIRREL

The tufted-ear squirrels are too unusual and limited in range for most of us ever to make their acquaintance, but there are, fortunately, other species, perhaps not as handsome, but equally entertaining and well-distributed.

One of these is the fox squirrel, *Sciurus niger*. The largest of all our North American species, the fox squirrel formerly was found in considerable numbers through the woodlands of all the eastern country. In numerous subspecies, varying considerably in size and color according to locality, from the rusty-yellow form of the Middle West to an almost black form in the swamps of the lower Mississippi, it is still abundant enough to escape extermination if properly protected. In body the fox squirrel is heavy and large and tends to fat in the fall. Because of its large size and tastiness, it has been the target of generations of red and white pot-hunters, and a real danger of extinction threatens certain of the several subspecies.

The very dark individuals of the South are the best-looking and most distinctive in appearance. White nose, ears and feet give them a comical look of just having been caught at the flour barrel. Fox squirrels have never been as abundant as the grays and apparently lack the latter's adaptability.

Being large and more difficult to handle, the fox squirrel is less likely to be molested by certain of the enemies which continually menace his smaller kin. This comparative freedom has made him a little lazier. Although definitely a tree resident, the fox squirrel spends a good deal of time on the ground, searching out food and storing it away for winter use. In disposition it is less excitable than other squirrels and more hardy in captivity.

GROUND SQUIRRELS

To those brought up in western agricultural regions, the term "ground squirrel" is not one to convey a favorable impression. The continued inroads of these rodents upon orchard and farm and their rapid increase in number under good conditions have alone been enough to make them thoroughly unpopular. Add to this the bitterness engendered in the rancher's heart when a favorite horse breaks a leg in a squirrel hole, and you have a fairly conclusive case against them.

Nevertheless, as a child my own first pet was a California ground squirrel, whose misfortune it had been to have the family burrow flooded by an irrigation ditch break. He was caught by a well-intentioned old farmer, put in a rubber boot for safe keeping and brought in as a present for me. It was my first serious naturalistic problem and I botched it. Reaching into the boot—it took a sight of stretching for an arm so short— I took hold of the only part of the squirrel I could feel, the tail, and pulled. It came out—but not the squirrel. There was nothing for me to do but squall and, helplessly holding that pathetic tag of gray-brown fur, I did. Fortunately, John Eaton was the original Mr. Fixit, and his gentle old hand, not without suffering a nip or two, got the squirrel out of the boot. There was a glint of knife and the twitching and bare vertebrae which my jerk had left all naked and exposed, disappeared. Put in a box, the squirrel sat up and washed his face. Apparently the docking had affected neither his feelings nor his dignity.

Inevitably, we called him Bob and, for some weeks he was gentle, quiet and lacking in aggressive qualities. With growth, he became wilder and lost his willingness to be handled. After biting everyone in the neighborhood, including a retired professional bad man and the minister's wife, he dove out of the cage at feeding time one morning and was last seen going over a ditch bank, headed for the wild life and self-expression.

This seems to be their general pattern of behavior in captivity, although cases have been reported in which California ground squirrels have retained the gentleness of infancy in maturity.

The habit of many mammals is to sleep through the cold winter months when food is scarce and the outer world glazed with ice and snow. The ground squirrels, in the arid and dry parts of their range are confronted instead with the problem of living over long periods of drought and heat. Their solution is the same as that of their cold country brethren. Deep in their burrows they sleep until the rains of fall have again brought green food and relief from heat. This sleep is called "aestivation" to distinguish it from the cold climate "hibernation."

The enormous group of species of ground squirrels are lumped under the general term, *spermophile*, meaning seed-loving. That the name fits is one of the farmer's greatest complaints, and thousands of dollars are spent annually in the control of these pests.

HARRIS GROUND SQUIRREL

SIERRA MANTLED GROUND SQUIRREL

THIRTEEN-LINED GROUND SQUIRREL

National Park Service photo by Joseph S. Dixon

CHIPMUNK

U.S. Fish and Wildlife Service, photo by Peter J. Van Huizen

FLYING SQUIRREL

THE ANTELOPE GROUND SQUIRREL

Of the several genera and many species and subspecies lumped under this term, only two groups are sufficiently attractive in disposition, appearance and habit to be worth our consideration. The first of these is the antelope ground squirrel, *Citellus leucurus,* which in several very similar subspecies ranges throughout the Southwest and southern California. The popular name of this squirrel has nothing to do with speed. A habit of running with the tail curved over the back so as to show the white undersurface, thus producing an effect somewhat like the white rump patch of the fleeing antelope, is the origin of the name. Much smaller than the other spermophiles, the antelope squirrel is a resident of arid and barren plains and slopes.

In captivity it has proved sufficiently attractive and adaptable to warrant pet shop dealers in southern California cities carrying it in stock.

THE GOLDEN CHIPMUNK

Under a number of local names such as golden chipmunk, calico chipmunk, mantled ground squirrel and big chipmunk, the spermophiles known scientifically under the name of *Spermophilus lateralis* and its many varying regional subspecies are one of the most familiar and popular creatures of the western woodland. Although true ground squirrels, it is easy to see why these are called chipmunks. With striped sides and rusty head and shoulders, they have the look of heavy set and slightly corpulent chipmunks. The likeness is not only one of coloring and pattern, but also one of habit and behavior. They frequent the same localities as the tiny western chipmunks and are commonly seen playing, working or begging side by side with their small cousins. Like the chipmunks and unlike the tree squirrels, these ground dwellers hibernate, and are only seen outside their burrows in very mild winter weather.

The golden chipmunk prefers partly open country: the edges of heavy forest areas, or partly logged or burned-over ground with plenty of stumps and boulders for sentry posts. Their nests are usually self-made burrows beneath logs, stumps or rocks, but they frequently take up quarters in comfortable and dry holes beneath abandoned cabins and buildings. The golden chipmunk is one of the reliable sources of fun for the visitor to the western National Parks. In Yellowstone, Yosemite and Estes alike, they are the first wild animal to welcome the tourist and ask him gently if he can't see his way clear to break out a bag of peanuts as a gesture of friendship. The visitor is usually delighted to comply. The inevitable result is that the squirrels around much frequented camp sites become so fat they fairly waddle.

As might be expected, the golden chipmunk does well in captivity and, if reasonably well cared for, lives out a goodly term of years. Its habit of hibernation in winter should not be interfered with, but catered to by seeing that it has a proper place to sleep and that it is provided during the late summer and fall with food of a type suitable for storage; nuts, seeds, grain and the like.

THE CHIPMUNK

Storage of food for winter use is a habit of practically all squirrels. Perhaps it is no more pronounced in the true chipmunks than in some of the larger squirrels; nevertheless in the portioning out of scientific names, it was decided to make the names of the two American genera of chipmunks, Tamias and Eutamias, a recognition of this storing trait. Tamias is from the Greek and means a steward or one who lays up stores. Anyone who has seen a chipmunk pack an incredible bulk of seeds into his cheekpouches and return again and again for more as long as the supply lasted will concede that the name is apt and descriptive. The intensity with which they tackle the problem of winter provision almost makes one believe that somewhere in the chipmunk mind each fall is a firmly fixed idea that spring will never show green leaves again.

Fortunately, industry hasn't much to do with sociability and although always busy, the chipmunk is willing to be friendly and accepts any assistance in adding to the winter store or satisfying present hunger.

The eastern chipmunk, *Tamias striatus,* in its several regional subspecies, is an almost constant companion to those who walk in the wood and farm lands. Much smaller than the red squirrels flitting about the trees of the same localities, Tamias is easily recognized by the black bordered light bands along each side and the dark stripe down the back. A ground dweller, seldom climbing far into the trees, the chipmunk lives in well-concealed burrows under old stumps, roots or rocks. The establishment is sometimes extensive and consists of several connected chambers for food storage and living. Exits for emergency use are provided. At the lowest level, toilet facilities are found.

The chipmunk day begins early. Before the sun is up and while the calls of the birds are still mere sleepy twitterings, he is about. Around camp sites, it is his usual morning practice to make a thorough check of the premises before the owners are awake and divert to his own use any stray bits of palatable and portable food left out. These informal and, as it were, unofficial calls do not prevent him from paying his respects at intervals during the day. If at all encouraged, he is likely to become practically a member of the party and often comes to regard the place as his own. When several chipmunks get the same idea about the same place, the day becomes a succession of quick and decisive rough-and-tumble scraps between the various claimants to the title of camp boss.

In the western genus of chipmunk, Eutamias, the wide range through an extremely varied series of habitats has produced a correspondingly varied series of species and subspecies. Although similar in appearance and general habit to the eastern chipmunks of the genus Tamias, these Westerners are, with the exception perhaps of the dark-colored forms from the northwest woods, smaller in size than their eastern relatives. The desert species, with less protection in the form of vegetation from a horde of enemies—hawks, coyotes, snakes and the like—are shy and wary and take cover at the slightest alarm.

All of the northern chipmunks hibernate, but their sleep is not continuous. A few days of warm weather brings them out to see what's going on, and they seldom wait for the melting of the snow to appear in the spring.

THE FLYING SQUIRREL

Of all our small animals, none are more attractive than the flying squirrels of the genus Glaucomys. As the popular name indicates, this squirrel is one of the few mammals to attempt the conquest of the air. Although unable to fly in the sense that bats and birds fly, these squirrels have the ability to volplane through the air for considerable distances. Long folds of skin along the side, extending from wrist to ankle, stretch out when the squirrel spreads its four limbs and leaps from a tree. Supported by these, the squirrel glides in a long graceful swoop to a lower objective. The wide flattened tail apparently acts as a steering mechanism and by this means, the squirrel is able to control its flight to some degree. Although this control is not much, it is enough to keep the squirrel from dashing its brains out and to enable it to make a four-point landing at the spot desired.

Flying squirrels make their homes throughout almost all the woodland areas of the United States, but are one of the least familiar of animals, even in regions of their greatest abundance. They are strictly nocturnal in habit, and except in an emergency, seldom venture out in the daylight hours. The result is that their presence within a stretch of woods is often unsuspected by persons frequenting it regularly.

Commonly living in holes in trees, the flying squirrels sometimes make nests of leaves and bark high in the branches if there is a shortage of suitable apartments. Often such a nest has as its beginnings the abandoned nest of a bird or of another squirrel. Man's inventions and buildings are not scorned. Many a bird box, vacated in fall by its legitimate residents, is taken over and occupied all winter by flying squirrels, very often with no one the wiser. Occasionally, they effect an entrance through a woodpecker hole into the attic of a summer home and the owner, opening up the house in the spring, finds the premises none the better for their occupancy. No more destructive than other rodents, a community of flying squirrels can, nevertheless, by converting bedding, drapery and the like to a form suitable for their purposes, make a house look as if it had been the winter haven of all the wild life of the region.

For flying squirrels, unlike other squirrels, tend to live in groups. These communities, accidentally perhaps, have stores in common, and generally speaking get along with one another in a quiet and amicable fashion quite unlike the continual feuding of the chickarees.

Big-eyed, gentle and quiet, and with soft, silky fur, the flying squirrels, more than any other, engage the interest and affection of people fortunate enough to know them either in nature or captivity. Extremely timid and shy, instead of attempting violent escape by means of tooth and claw when roughly handled, they are likely to become so terror-stricken and paralyzed with fear as to die from the shock.

Because of nocturnal habits, the flying squirrels, as show animals, do not furnish the amusement some of their relatives do. As pets their beauty and gentleness and the evident liking and recognition they speedily give to considerate owners, overbalances all other considerations and ranks them first among squirrels in the naturalist's affection.

Their food in nature is most varied. Seeds and grains of all kinds, fruits, nuts and acorns, the buds of trees and shrubs, insects and their larvae and such meat scraps as come their way are all fare to the one table.

Normal squirrel and chipmunk diet varies throughout the year and in captivity, may be varied likewise. Acorns, nuts of all sorts, rolled oats, barley, dog biscuit, apples, carrots and such fruits and vegetables as are found acceptable, should be occasionally supplemented with an egg or a bone with shreds of meat hanging to it. Feed daily in the late afternoon. If much food is left over in the morning, cut down the quantity.

The most suitable cage is a large outdoor pen, completely enclosed with wire mesh. With a sheltered nesting box or hollow tree trunk, sunshine and protection from wet winds, a pair of squirrels can be left out all winter and may possibly present the owner with a fuzzy little family in the spring.

Small Rodents and Shrews

THE MANY species of small rodents native to North America offer infinite possibilities for study and observation. Under the terms "rat" and "mouse," with their inevitable hint of destructiveness, filth and plague, a number of species easy to obtain and attractive as pets have been avoided. Our native rats and mice are not close relatives of those nuisances, the house rat and the house mouse. These latter originated somewhere in Asia and, along with other blessings of civilization, spread with commerce to every part of the globe. Their extreme adaptability and fertility have made them a serious menace to the health and welfare of peoples everywhere.

Of those several pests, the brown rat, *Rattus norvegicus,* called also the Norway rat, is particularly voracious and destructive. Its great numbers, filthy, disease-spreading habits and omnivorousness have made it the foremost animal pest in existence. Its migrations from the Old World to the New were preceded by similar migrations of the black rat, *Rattus rattus,* a smaller and less objectionable creature. Their inevitable conflict in range and competition for food have resulted in an almost complete elimination of the black rat from the United States.

Indeed, the black rat's survival in this battle of the species has been largely as a creature of captivity. In the albino and particolored strains developed from the original stock by breeding, the black rat is in constant use in laboratories and hospitals and has been of inestimable value as an instrument of medical research. Its gentleness and responsiveness have made it, in its varieties, the rat of the pet shop and the nature room. Among the common types available are the pure white, the black, the black-hooded, the red (a light auburn shade) and the red-hooded. They breed fast and if handled from infancy seldom turn vicious. Even very old males, coarse-haired and tough-looking, dispensers of heavy-handed discipline in dealing with the affairs of a family cage, often display an unexpected gentleness and a liking for being scratched. Youngsters are, as might be expected, full of curiosity and have an enviable talent for good-humored, rough-and-tumble play. An English walnut, tossed into a cage of young rats, immediately starts a riotous free-for-all football game.

THE WHITE-FOOTED MOUSE

Of our native rodents, the one most widely distributed and most likely to find its way into the traps of the collector is the white-footed mouse, *Peromyscus leucopus*, or one of its very close relatives. Thoroughly adaptable, this group ranges through most of North America in a series of species and subspecies, each suited to the particular environment.

As their name indicates, the white-footed mice have pure white feet and underparts. The body color is the brown or brownish-gray often described as "fawn" color. This coloring has given rise to another popular name—the "deer" mouse.

The white-foot is small, a little larger than the house mouse, however, and very quick and light on its feet. It is an agile climber and jumper. Found in both forest and field, it forages about inpartially for seeds of grass and plants, nuts, insects and such bits of stray meat as may fall to its portion.

Even in its haunts of the far North, the white-foot disdains the comforting refuge of hibernation but, from burrows beneath logs, from nests in the hollows of trees or in the crevices of rock, sallies forth to forage and see the winter world. Its characteristic tracks are often one of the few signs of life to be seen in the snow-covered woods. Extremely provident, the white-foot takes no chances with bad weather but, in late summer and fall, stores up an ample supply of provisions to last over until spring.

The white-footed mice are gentle toward humans, and they are trusting if their growing confidence is not destroyed by some untoward event. In camp, they frequently make a regular habit of visiting and looking about for such bits of food as are to be had. If encouraged and fed, they practically take possession and, unless food and supplies are kept in mouse-proof containers, can be quite a nuisance. Under these wild conditions they have small fear of people and readily learn to take food from one's hand.

The behavior of the white-foot toward human beings has, unfortunately, nothing to do with its attitude toward its own kind. Even in the wild they are pugnacious with one another and resentful of raids upon chosen food territories. This seeming greediness is really an attempt to make hay while the sun shines and to take away in thier bulging cheek pouches as much provision for winter storage as is possible.

In captivity, this aggressiveness toward one another makes it unwise to keep too many in the same cage. In settling their problems of precedence and territory, these inoffensive-looking little creatures display a formidable courage and ferocity and death is the ultimate referee in their disputes. Extremely active in nature, the white-foot when caged often gives the impression of having gone completely and suddenly mad. The gyrations and crazy circuits of the cage which give this impression are but a substitute for its normal travels in the woods and a means of keeping in shape for an emergency.

Persons fortunate enough to witness and hear an exhibition of one of the celebrated "singing mice" have sometimes assumed that the tiny songster was seeking by music to enter into a lady's affections. In these singing mice, different sorts of sounds have been described as the song, but ordinarily it is spoken of as a bird-like twitter, weak in volume, variable in pitch and intensity of tone. Many species of mouse, from all parts of the world—the house mouse particularly—have been reported as singing. The white-footed mouse tribe has been known to produce vocalists.

The most plausible explanation of these musical prodigies in mousedom is based on the known ability of many animals to hear sounds high above the range audible to the human ear. A whistle on the market takes advantage of this quality of the dog's sensitive hearing. The whistle is pitched above the hearing range of the human beings about and does not disturb or attract their attention but is plainly heard by the dog it is intended to recall.

In the case of the singing mice, it is not unlikely that most species have a normal song or twittering above the range of our dull ears. In exceptional cases, the pitch is sufficiently lowered for the song to become audible to us and—the mouse sings!

THE MEADOW MOUSE

With the exception of the more arid regions, the field or meadow mouse, *Microtus pennsylvanicus,* is to be found throughout the United States in some one of its forms. A little effort will reward the collector by the addition of a few of these inoffensive creatures to his list of animal guests. As individuals, they are really inoffensive. In nature, unfortunately, they never seem to exist as individuals but almost always as great numbers and constitute a serious agricultural problem. In spite of the incessant warfare waged upon them by the hawks and owls, the fox, bobcat, snakes, and even by some of their own kin, the field mice are so prolific that in infested regions crops have been laid waste by them and serious permanent damage done to orchards through the nibbling of tree trunks. This latter damage usually occurs in winter under the cover of concealing snow. Unlike the provident white-footed mice, few of the field mice lay in a supply of winter food. Active during the winter, to keep alive they are put to the extreme measures of devouring bark and other plant materials not bothered in summer.

In summer, they feed upon almost anything: growing grass, alfalfa and grain; seeds and nuts; bulbs and tender roots. The orchard, the flower garden and the vegetable plot supply fare for the satisfaction of their demands.

Their presence in a locality is evident in the inch-wide runways, little well-traveled roads, meandering through the grasses and weeds near their feeding grounds. These roads lead to the entrance burrows of the warm and dry grass-padded, subterranean nests, and are religiously kept clear of debris. Bits of stick, twigs and pieces of grass are promptly removed lest they impede by the fraction of a second a homeward flight from disaster.

Several species of field mice may be found within the same territory. Close observation will show that, although the ranges seemingly coincide, in reality they do not. Each has its own habitat preference. One prefers the high, dry ground; another the grassy meadows. Still another may be found paddling around in the marshes and on the borders of the streams, thoroughly at home in the damp. However, to the normal hazards of a field mouse life, the latter must add the appetites of the bull frog and of the larger fish, for to escape danger, it is often forced to take refuge in the water.

THE RED-BACKED MOUSE

The red-backed mice of the genus Clethrionomys, close relatives of the field mice, are of about the same size but have somewhat brighter colors. Unlike the field mice, the red-back comes into but little conflict and competition with man in its forested and primitive mountain areas. Occurring in greatest numbers in the cold regions of the north, its range follows the mountain ranges down as far as North Carolina in the East and the lower Sierra Nevada in the West.

THE LONG-TAILED TREE MOUSE

Another reddish-colored mouse is the long-tailed tree mouse of the Pacific Northwest, *Phenacomys longicaudus*. This little animal, physically resembling the ground-dwelling field mice, has become largely arboreal in habit and lives high in the branches of great forest trees. Shy, nocturnal little creatures, these mice are seldom seen and often the only evidence of their presence is the bulk of one of their clumsy-looking and elaborate nests high in a tree.

Their food habits, because of their environment, are quite unlike those of other mice. Instead of eating anything that comes to hand, the tree mice feed upon the tender parts of the needles of the conifers in which they make their homes. They are an example of a native animal, apparently fairly common within its range, with gaps in its life history which the observations of some interested amateur could fill.

THE GRASSHOPPER MICE

This group of dainty small mice of the semi-arid western highlands is of considerable interest because, unlike most mice, instead of being an insidious enemy to man, it is a friend and ally.

In two species, *Onychomys leucogaster* and *O. torridus* and their many regional subspecies, the grasshopper mice wage a continual and unremitting warfare against the destructive insect world. They are not entirely insect-eating, but include in their fare scorpions, caterpillars, other species of mice and similar small game. When living food is not available, they subsist on the seeds of weeds and grass and other vegetable material. Their range is through the arid plains of the West and Southwest, regions of open and comparatively dry winters, which make any thought of hibernation unnecessary. However, winter food does become short and the grasshopper mice guard against mischance and hunger by storing away seeds and nuts during plenty. They make their homes in burrows in the ground.

The grasshopper mice are rather stocky in build, soft-furred and clean-looking. About the size of the white-footed mice, and, like them, having white feet and white underparts, they may be distinguished from the latter by a heavier build and shorter tail.

Their senses of smell and hearing are very acute and the slightest rustle of an insect in the vicinity or the scent of living prey is quickly caught up and followed to a logical conclusion.

Less nervous and timid than other mice, they display no great fear of man and quickly become accustomed to the presence of human beings. Some individuals seem not to mind handling and from the time of capture, display no dislike for such attentions. Others resent being bothered, and, if necessary, use their teeth in protest.

In his paper on these mice Mr. Vernon Bailey of the Department of Agriculture describes them as not being fit for children's pets, ". . . but they will rid kitchens, basements, cellars or greenhouses of cockroaches and other pests. When so used the mice are easily handled and controlled merely by placing their cages with open doors in the room and allowing them to run at large at night. Almost invariably they will be found in their own nests in the morning."

THE POCKET GOPHER

The possession of exceedingly useful and large cheek pouches, lined with hair and opening outside the mouth at the lower edges of the cheeks, distinguishes three related groups of our North American rodents from all others.

The pocket gophers, burrowers in the dark, vicious of disposition, are of considerable interest to the farmer because of their damage to crops and trees. Unsuitable for captivity because of habit and character, we can only pass on with wonder from this surly group to their relatives, the pocket mice and the kangaroo rats. In contrast to their churlish heavy cousin, these rodents are gentle in behavior, pleasing in color and delicately formed.

THE POCKET MICE

The pocket mice of the genus Perognathus vary in size from the tiny yellow species, *flavus*, one of the world's tiniest mammals, to a number of species of about the size of the common house mouse. Their kinship to the gopher is evident in the carrying pouches of the cheeks. The resemblance stops there. With their long tails and feet and legs adapted to jumping, they are much more like their larger cousins, the kangaroo rats.

Both the pocket mice and kangaroo rats are peculiar among animals in that they seem never to drink water. Desert dwellers of the arid Southwest, they have gone far beyond the camel's simple trick of making a traveling tank of himself and apparently have dispensed with dependence upon natural sources for water. There is evidence to support the claim that through chemical changes during digestion the dry starches contained in seeds yield enough fluid to support the life processes.

Within their range, the pocket mice are far more common than the day traveler suspects. Aside from their tiny tracks and the small mounds of earth marking the underground burrows, there is no evidence of them. At night, in favored localities, they appear in great numbers and, given a little encouragement, cheerfully invade a camp and make themselves at home. So much at home, indeed, that family squabbles over bits of food put out for their entertainment are the rule and not the exception.

In disposition, the pocket mice are gentle, easily tamed and cared for. Their nocturnal habit is a disadvantage in some cases; an advantage in others. For city dwellers who are away all day, animals that do nothing but sleep during the evening are the wrong pets.

THE KANGAROO RATS

The kangaroo rats of the many species and subspecies of the genus Dipodomys share the range of their cousins, the pocket mice. Their popular name comes from their miniature kangaroo-like form. Long hind legs for jumping, long tail and the habit of carrying the weak forelegs hanging ineffectually in front of the body, give a kangaroo impression. In size, most of the kangaroo rats are about that of a small rat or large mouse. The tail is not fuzzy, but well-haired and slightly tufted on the end. The fur is silky and long, of a buff shade lightly sprinkled with black hairs on the upper parts. In most species black markings occur on the ears and tail and a conspicuous broad white line cuts the body color across the flanks. Like the pocket mice, the kangaroo rat has cheek pouches for carrying food. These, when empty, are not unduly noticeable. When full—it takes an unbelievable amount of seeds to fill them—they impart to the animal a definitely serious-minded look.

The kangaroo rat shows decided cleverness in the storage of food. Instead of taking seeds immediately to the permanent cache down in the burrow, it places them in shallow pits above

U.S. Fish and Wildlife Service

KANGAROO RAT

National Park Service photo by E. R. Warren

WOOD RAT

John C. Orth

MEADOW MOUSE

American Museum of Natural History

WHITE-FOOTED MOUSE

D. Dwight Davis

SHREW

ground and covers them lightly with sand. The heat of several days' desert sun drives off all moisture, and seeds which would have become moldy and spoiled had they been cached green, are effectually cured and stand an indeterminate period of storage.

The largest of the kangaroo rats, *Dipodomys spectabilis*, is especially attractive in captivity. About a foot in total length, a great deal of which is the tufted tail, these animals are large enough to have under foot without too much danger of an accident. Their gentleness, cleanliness and lack of destructive tendencies make it possible to permit them the run of any cat-free place. Given such freedom in a house or apartment, the kangaroo rat will make himself comfortably and respectably at home. He has a complete trust in people as friends, but never entirely believes in them as safe providers. Old habit leads him to stow away seeds and nuts in all sorts of odd places, as provision for a threatened hungry day.

In nature, the kangaroo rats practice a trick which they occasionally carry over into captivity. Confronted by an unfamiliar object, the rat kicks sand at it in an attempt—usually successful —to find out whether or not it lives and is menacing.

THE JUMPING MICE

Somewhat like the kangaroo rat in physical character but actually not very closely related, are the jumping mice of the genera Zapus and Napaeozapus. These really are allied to the Old World group of kangaroo ratlike rodents called jerboas. A long tail for balancing, large hind legs and feet enable the jumping mouse to make unbelievably long leaps.

During the summer jumping mice are often seen in the meadows and woodland of the northern half of the United States and in Canada, up to Hudson Bay. In winter they hibernate, but, like the chipmunks, occasionally emerge on a fine winter day. These mice are inoffensive little creatures and even when first caught—it takes fast work to catch them by hand—offer little resistance and seldom attempt to bite. The young eagerly and willingly attempt to adapt themselves to drinking milk from a spoon instead of a natural source, but in spite of this willingness they are difficult to raise.

THE WOOD RATS

One genus of American rodents, Neotoma, has, through a

rather peculiar habit, gained a reputation for doubtful honesty and practical joking.

These, the western trade rat or pack rat and its eastern relatives, the wood rats, most often make their visits to us known by their "trades." Small things disappear, bright articles especially, and in their place is left a chip, a bit of cactus, a piece of weather-worn bone, a horse dropping or some other equally absurd—and from the human point of view worthless—bit of debris. The rat doubtless sets some sort of valuation upon the object left in exchange for the thing which has caught his fancy, but there is no idea in his mind that he is making an exchange. It is merely that on his nocturnal rounds, he has picked up a bit of trash to add to the pile forming his nest or littering the ground about it and, later coming across something which pleases him better, abruptly and without benefit of bargaining, drops his original *objet d'art* and trots off with the second selection.

In the East, wood rats are present in most of the mountain areas but are not common, and their presence in a locality is often unsuspected by the human inhabitants. It is in the West that the tribe and their activity are most evident, most amusing and most disconcerting. There is hardly a cabin in the western mountains that is not visited from time to time—usually nightly —by one or more of these engaging rodents. If there is any way of getting inside, the fun begins. There is little judgment exercised in the pack rat's treasure hunting. He does seem to have, however, an uncanny knowledge of what you can least afford to lose, and a perverted sense of humor in what he chooses to leave in its place. When my fountain pen disappeared from a cabin in the Coast Range, I could not feel that the ancient rabbit bone left was adequate pay. Later, when part of my safety razor disappeared and a piece of twisted and dry cactus took its place, I decided to do a little trading myself and went on a tour of such nests as I could locate. Frequently, lost articles may be reclaimed by such hunts and, in this case, I was lucky. Unfortunately, with some species and localities, the nests are so large, so compact and bristling with cactus, as to defy anything but whole-hearted and reckless investigation. Often pack rats decide that the cellar of a house or the space beneath the floors offers a suitable and well-protected home site. Having discovered such a space and decided to build in it, they labor industriously to fill it completely with anything and everything.

Although called rats, the wood and pack rats are no more like the filthy and disease-bearing rightful owner of the name than are the squirrels and chipmunks. Indeed, the brushy-tailed pack rat, *Neotoma cinerea*, with its bushy, well-haired tail and white underparts, seems much more like a chipmunk than a rat. None of the wood rat group has a naked and scaly tail; in all the species the tail is well covered with hair and the general impression of the animal is that of a clean, inquisitive and intelligent creature with no harm in him.

As in other animals, individual temperament and species (as well as the tactics of the keeper) have a great deal to do with the behavior of wood or pack rats in captivity. Some will be found inquisitive and friendly; others wild and lacking in trust. All, ordinarily, adjust themselves to the rigors of cage life fairly well, and, if given roomy and dry quarters, an adequate food supply and reasonable freedom from disturbance until they are accustomed to their new home, pay their way in furnishing amusement and instruction.

SHREWS

One of the greatest satisfactions vouched the animal keeper is to be able to maintain in health, species of delicacy or rarity. These may offer nothing in the way of amusement or affection —indeed, may be definitely unpleasant in behavior—but the life history notes and knowledge made possible by successful captive maintenance are well worth the trouble. The shrews are creatures of this sort. Very small, difficult to trap without fatal shock or injury and hard to keep, only a few of the many species of this large group of little savages have been observed for any length of time in captivity. In nature, their habits are such as to make the recorded observations of their behavior sketchy and haphazard, and there are many gaps in our knowledge of their life histories which only patience and interest can fill.

The shrews are the tiniest of mammals but have a courage and ferocity entirely disproportionate to their bulk. They are fierce and bloodthirsty and, without a sign of hesitation, will attack and swiftly kill creatures many times their own size. They are such creatures of blood that if size were left out of consideration, the killing exploits of those notorious butchers, the weasel and the wolverine, would seem relatively tame by comparison.

The fact lying behind the shrew's viciousness and incessant hunting is its never-resting hunger and need for food. The life processes of these tiny animals are unbelievably swift; a high metabolic rate and great activity demand large amounts of food and fasts of more than a few hours result in the shrew's death.

As might be expected, vegetable food, although forming a part of the shrew diet list (especially in emergencies), is hardly concentrated enough to suit the needs of this incarnate appetite, and in nature it subsists for the greater part upon insects and their larvae—cutworms, tent caterpillars, slugs, moths, bees, flies, grasshoppers, crickets, and even larger animals such as mice. Probably also, the nests of rats are at times invaded and the young butchered and devoured.

The group is large and widespread over most of North America. The genus Sorex, in many species and subspecies, has almost continental distribution. It is generally called the long-tailed shrew to distinguish it from its short-tailed relatives of the eastern coast, of the genus Blarina. The shrews are like small mice in size and appearance, but do not belong to the rodent order. Instead, with their close relatives, the moles, they make up the order of Insectivora or insect-eaters.

The several species of both Blarina and Sorex are forest dwellers, and their rustling among the leaves bedding the forest floor has caused many a stroller to stop in mid-stride and try to trace the source of that dry and scuffling little noise.

Shrews are primarily creatures of the night and the dark hours mask their greatest activity, but they are frequently about during the day, especially in deep woods.

In captivity, it is not only necessary to attempt to duplicate as well as possible the normal shrew menu, but it is also imperative to provide certain conditions of habitat. They are accustomed to living in damp moss and earth; and when placed, as they often are, in clean and dry cages of a type suitable for rodents, captive shrews quickly succumb, not to starvation, as is ordinarily supposed, but to a lack of humidity and too high a temperature. The cool woodland terrarium, described on page 85, with its growing plant materials and mossy hummocks, provides the cool humidity and the opportunities for concealment so necessary to shrew well-being. Shrew food habits are not such as to destroy or damage the plant life of the terrarium, and in its shelter these tiny furred creatures may truly feel at home. A small nesting box or a tin can, half-filled with earth and dry moss, and partly buried in the earth in a far corner, makes a satisfactory sleeping chamber. It may be covered and masked by a blanket of moss, a piece of root, bark or other suitable and fitting accessory.

Unless the container is very large and the food supply more than ample, one shrew to a cage is the limit, for their savagery does not stop at their own species. It is obvious that shrew-mating in captivity is not an easy matter. Before a compatible pair is found blood is likely to flow.

In addition to the regular supply of whole foods—meal worms, earthworms, grasshoppers, crickets, all sorts of insects and their larvae, and mice or rats—vegetable foods such as sunflower and other seeds should be kept always accessible. If insects and small whole mammal carcasses are not to be had, liver and other meats may be substituted for a time.

In addition to the members of the genera Blarina and Sorex, the shrew Neosorex, commonly called the water shrew or marsh shrew, is found in the northern part of the United States. A number of species and subspecies are known, but information on their life histories is still incomplete. Aquatic in habit and always found in the vicinity of marshes and streams, these shrews in captivity would certainly need the conditions described in the marsh-stream terrarium (see page 87) and in such a set-up might reasonably be expected to survive. Their food in part comes from the water—small frogs, water insects and the like.

Larger Mammals

THE SMALL warm-blooded creatures, while relatively easy to maintain, cannot be expected to compete for human favor as pets with some of the larger forms.

THE RACCOON

Most widely known and popular of these is the raccoon, *Procyon lotor*, and its many subspecies. Of all our wild life, the raccoon is perhaps the most distinctive. Existing nowhere except in North America, and associated in our minds with such wearers of coonskin caps as Daniel Boone, Crockett, Audubon and a score more of romantic pioneers, the 'coon seems American indeed.

The raccoons, although ranging through the northern United States and into Canada, are not cold climate animals. Their widest distribution is through the southern states and the milder parts of the west coast. Generally they avoid high elevations and are found along the wooded shores of streams and lakes and in from the water and their footprints in the wet sand or clay of stream or lake margin betray their presence in a locality. The 'coon nest is ordinarily within less than a mile of the water, commonly in a hollow tree, and sometimes quite high above the ground. A shortage of hollow tree apartments has been known to reduce 'coons to more humble and less safe quarters in the abandoned burrows of other animals and in large rock crevices. One reason for the hollow tree preference is that the raccoons, although night hunters and travelers, like to roost in the crotch of a limb handy to the home hole and take a sun-bath. Climbing is one of the 'coon specialties and while not a swift runner, he is able to get over the ground. When pursued—and being pursued is not an uncommon happening in the life of a 'coon —they exercise a considerable intelligence, and are able to make quick decisions to fit their tactics to a situation outside their former experience.

For his size, the raccoon possesses great strength and knows how to use it. Treed by a pack of dogs, a 'coon has often been known to drop down into the middle of the howling melee and by cutting and slashing tactics, fight his way out. Raccoons are less solitary than most other woods creatures and the mother keeps the young with her, teaching them, and often going long distances in search of food, for nearly a year after their birth in the spring. The male parent is more independent and assumes no responsibility in the matter.

The food of 'coons is extremely varied in the wild and, in addition to freshwater food such as clams, crawfish, frogs and turtles, neighboring fields are robbed of corn, vegetables, nuts and fruits. Birds and their eggs and, if available, poultry, are on the diet list. His liking for the latter renders the 'coon somewhat unpopular in farming circles. Obviously, with such an appetite, 'coons are not difficult to feed in captivity. Table scraps, well-varied—meat, fish, bread, vegetables and eggs—provide a good ration. The 'coon is inclined to gluttony and his excessive appetite should be kept in check by feedings a little scant of his desires. Otherwise, his waistline and temper suffer.

A habit of the 'coon in captivity is to wash eatables in water before devouring. Held in the front paws—these are very efficient hands—the food is usually at least rinsed. In the wild, this habit probably does not exist except when muddy clams, crayfish or frogs are being eaten. In captivity, the water is convenient and almost everything is given a souse.

Even lump sugar and candy are not exempt from washing. One 'coon of my acquaintance, clever enough to go through pockets in search of sugar, invariably dunked it in his water pan. The dunking naturally was followed by the sugar's dissolution and a worried and bewildered expression on the 'coon's face. Hurried gobbling and unmannerly licking of paws quite unlike the 'coon's usual dainty feeding served to salvage some of the mess.

'Coons are extraordinarily handsome animals, with their ringed tails and masked faces. As in the bear, the hindfeet rest flat on the ground and the track is much like that of a small child.

In captivity, older 'coons resent attempts at gentling and very seldom become really tame. Cubs, taken from the hollow tree nest, adapt readily to human society and soon become the favorite members of any household to which they are introduced. Their intelligence and liking for food, though, is too strong a combination to permit their being allowed loose in a house, for they are skilled at getting into mischief. On a chain, with a house to sleep in and a dead tree to roost in during the day, they are fairly happy. The ideal provision is a large cage of coarse wire mesh, four by twelve by six feet or larger, fitted with dead wood for climbing and sun-bathing. A small box house, windproof and rainproof, with a ladder leading up to it, may be put in an upper corner. The base of such an outdoor cage must be sunk into the ground. Although not to be classed as a burrowing animal, the 'coon has a way of getting out. If the pen is very large, an overhanging edge wide enough to prevent escape may take the place of a complete covering.

In such parts of its range as have really cold weather and heavy snow, the raccoon hibernates over the worst of the winter.

THE CACOMIXTLE

A very close relative of the raccoon—the cacomixtle, as the Aztecs named it—*Bassariscus astutus,* is found in the Southwest and West. Known by many confusing popular names such as the 'coon cat (many old timers profess to believe it represents a hybrid between a domestic cat and the raccoon) the cacomixtle has drifted north to us from its native Mexican home. In appearance, it resembles the raccoon in the possession of a long bushy tail with alternating light and dark rings, but is a slimmer animal. Like the 'coon, extremely intelligent and with a catholic taste in foods, it is easy to maintain in captivity. The miners of '49 in California found it an admirable substitute for the mousing domestic cats left in their eastern homes, and a 'coon cat was a resident of every well-appointed miner's cabin.

THE SKUNK

In sharp contrast to the popularity of the raccoons, the skunks enjoy no great favor. They are better known, perhaps, than the former, and their presence is obvious in a vicinity even to persons utterly without knowledge of natural history. The quality which makes them evident and unpopular is the possession of a unique and efficient protective device. This consists of two oval sacs, just under the skin below the tail. These sacs are filled with a heavy and pungent fluid, slightly phosphorescent, which, under provocation, is ejected with considerable force to a distance of as much as ten or more feet by means of twin ducts. Each sac contains but a few drops of the fluid and may be

Robert Snedigar

CACOMIXTLE

Courtesy American Museum of Natural History

EASTERN SKUNKS

U.S. Fish and Wildlife Service, photo by Robert W. Hines

WOODCHUCK

U.S. Fish and Wildlife Service, photo by L. K. Couch

PORCUPINE

emptied independently of the other. When in use they protrude from the anus and the skunk's own fur is not soiled by the discharge. The scent is for protection, not personal adornment.

As long as a skunk is head on, there is no danger. A quick reversal and elevation of the tail are indications that the immediate neighborhood is no longer healthy. Stamping of the feet is a preliminary to these more businesslike preparations. The little spotted skunk has a pose for threatening and for making good the threat that is not only unusual but downright theatrical. The alarmed animal does a handstand and, hindquarters and tail straight up in the air, peering backward through his front legs, moves—even rotates—to keep his opponent in view and range. A risky, but possible, method of catching a skunk, especially a young one that doesn't know his own strength, is to lift him quickly by the tail. One of the big drawbacks to this is the fact that sooner or later the skunk must be put down!

The skunk's spray is nauseating at close quarters, burning to the skin and, if landed in the eyes of man or beast, produces temporary and very painful blindness. On clothing, the odor is so persistent that it is difficult to eradicate it. Dogs, untreated after an encounter with a skunk, go about for months in a gradually decreasing aura of the scent. It may seem to have entirely disappeared, when a wetting will bring it out strongly again. Coats made of skunk skin, although subject to skillful furrier's deodorizing, occasionally in damp weather give out more than a hint of their origin.

There is a method of killing skunk odor which I have only seen demonstrated on an abject and ashamed cocker spaniel, but which obviously has other possible applications worth trying if the victim is human and clothing is involved. The dog was doused with canned tomato juice, which was well rubbed into the hair. A good rinsing was followed with a regular soapy bath. Doubters of the value of this method can be furnished with testimonials—even one from a most glamorous and beautiful TV star.

In spite of having originated and brought the principle of modern gas warfare to a high degree of efficiency, the skunk is by no means a pugnacious and vicious animal. Indeed, he never uses his weapon unless greatly provoked, and is calm and quiet under stress that would reduce most wild animals to a bad attack of nerves. He seems to feel that no matter what situation arises, he has at his disposal a sure means of controlling it. This self-confidence enables him to go almost anywhere he pleases and allows him to be insistent—even with man—on his trail rights. Taken unawares at digging for grubs or when prowling, the skunk's attitude is apt to be one of mild curiosity as to what *you* are going to do.

Economically, the skunks are among our most valuable animals, not only for their skins, but also because they render an important service to agriculture by their continual hunting for insects. Realization of this service is reflected in the increasingly stringent state regulations against their destruction. These restrictions have largely been sponsored and urged by farmers' associations in recognition of the animal's insect pest control value. The larvae of beetles, many of which are injurious, are one of the skunks' constant preys. Little pits in the grass of meadows mark their search for these underground pests.

In addition to a great variety of insect food, in nature the skunk feeds on almost anything—carrion, earthworms, berries and fruit, reptiles, small rodents and so on. One of its best food tricks is to roll toads in the grass to rid them of their poisonous exudate. In the same fashion hairy caterpillars are rendered edible.

In captivity, the skunk is gentle and easily cared for. Although it is not necessary to remove the scent glands in order to have him behave as a respectable and familiar member of the household, it is advisable. Skunks *have* been brought up intact without any unpleasantness, but accidents will happen. The removal of the scent glands in mature animals is a rather severe operation. In the young, however, it is relatively simple and without danger. In Farmer's Bulletin No. 587 the Department of Agriculture gives all the necessary instructions.

Like the raccoons, skunks appreciate and return human affection. If deodorized, they may in most cases be permitted much the same liberty as a cat. *Au naturel*, they are best taken care of in outdoor cages or pens. They are skilled burrowers and unless the sides of the pen are carried well below the surface of the ground they are not only likely, but almost certain, to dig out.

Their normal living quarters are in self-dug burrows, the abandoned holes of other animals or, in some instances, in rock cavities. In pens, these may be imitated by trenches covered over with plank and earth or by a barrel, cut in half lengthwise, and covered with soil. The den must be dry and in a location which will not be subject to flooding from rain.

Their food, like that of the 'coon, may be almost anything available. Feedings should be sparing enough so that no food

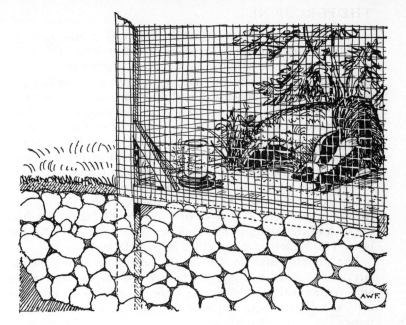

SKUNK PEN

is left lying about. Feed once daily in the late afternoon. If the skunks have bred and have young, meals of fresh meat should frequently be given to supply a desire for protein. If at all possible, a female about to bear should have a quiet pen of her own removed from possible disturbance. No attempt to see or handle the young should be made lest in her agitation over their safety, the mother kill them.

The female skunk keeps her young with her until they are in their second season, and it is not unusual, walking in the moonlight down a wood trail, to meet an old female with her family stringing single file behind her.

Our skunks belong to the two genera, Mephitis and Spilogale. The several species belonging to Mephitis are large black skunks marked with wide white strips down the back and tail, and range throughout most of North America. The Spilogale, better known as polecats or civet, are not quite the size of a small cat, and instead of striping have symmetrical markings of white on the black ground. These spotted skunks are principally distributed through the South and West.

THE PORCUPINE

The skunk is not our only small animal with a curious means of defense. The porcupines, protected by their long, stiff quills, are not to be lightly trifled with, as many a dog has found to his dismay and pain. In general movement, these heavy-bodied and squat animals are cumbersome and slow and, lacking the acuteness of most animals, fail to heed the astounding racket which marks human progress through the forest. Thus they are often caught unaware. When on the defensive, the quills, scattered through the fur, are erected by a contraction of the skin muscles; the porky bunches up, raises the brutal tail and is ready to flail at anything within reach. Contrary to general belief, the quills are *not* thrown. The best that the porcupine can do—and its best is very good indeed—is to strike the barbed points into the flesh of the aggressor by blows of the tail and shoves of the body. This he does with surprising force and speed.

The stupidity and slowness of the porcupine have been the salvation of more than one lost hunter and woodsman. Among all forest animals, the porcupine alone is one that an unarmed man, sick from hunger and exposure, can kill with a stick or a stone.

The porcupines are true rodents, as the marks of their heavy teeth on the bark of trees, hemlocks especially, attest. In addition to bark, they devour impartially almost anything in the way of vegetable material. Salt is their particular weakness and a hint of it is enough to set a porky gnawing endlessly. Shovel handles stained with human perspiration, barrels which have held salt meat, sweat-stained harness and saddles, all are destroyed for the meager taste of a little salt.

In captivity, this gnawing propensity makes it necessary to protect projecting parts and standards of the cage with metal. These animals are good climbers and a wide overhang or a complete top for the pen is essential. Green bark-covered sticks kept in the cage in liberal quantities tend to keep them occupied and give exercise for the teeth.

As lap pets, porkies are obviously impossible. Young ones tame easily and although handling is out of the question, learn to recognize their owner and make some response to his coming.

The porcupines, in two very similar species of the genus Erethizon, range throughout most of the forested areas of the United States and Canada.

THE WOODCHUCK

Although the woodchuck, *Marmota monax,* is familiar to every eastern farm boy and girl, these large, waddling rodents achieve popularity and make the front pages of the newspapers only once a year. On February second, Groundhog Day, the woodchuck or groundhog, whichever you please, supposedly rouses out of his hibernation and crawls to the entrance of his burrow for a look at the weather. The legend goes that if he is able to see his shadow—that is, if the day is sunny—the indications for good weather are not propitious and he goes back into his hole for a nap of six weeks' duration.

The woodchuck is really a large ground squirrel, living in self-dug burrows among rocks or in old walls. Heavy-bodied and compact-looking, short-tailed and bullet-headed, he takes no beauty prizes. His food is all sorts of vegetable material and he has a regrettable tendency to locate within convenient access of truck gardens in order to share the toothsome results of the farmer's labor.

The large and competent teeth of an adult woodchuck discourage the idea of taming one. Babies, however, fed milk by hand when young and later graduated to a diet of vegetables and mush, are docile and freely permit—even invite—handling. Many pet woodchucks appear in natural history literature, notably the one described by Audubon which slept beside the hearth fire and was on amicable terms with the family dog and cat.

Naturally, the woodchuck's food in captivity should approximate its wild fare. As to housing, the pen walls must be countersunk into the ground for several feet to foil attempts at burrowing. A plank- and earth-covered trench, secure from flooding, will provide the animal with the start of more complex and deeper diggings.

The western marmot, *Marmota caligata,* commonly called the hoary marmot, is the eastern woodchuck in a slightly larger edition, with certain revisions and alterations made necessary by a different climate and food supply. The hoary marmot is called in some regions the whistling marmot. In high altitudes where bird and animal life is scarce, the shrill piping of the marmot is often the only sound to be heard.

THE COTTONTAIL

Domesticated rabbits are easy to keep and their care offers no special problem. Rather lethargic in disposition, unexcitable and with good appetites, almost anything in the shape of a dry and warm shelter and a ration of hay, grain and root and green vegetables suffice for their needs. With the native wild hares, or rabbits (as we commonly and somewhat inaccurately call them) the problem of care is another matter.

Of the great number of cottontail rabbits of the species *Sylvilagus floridanus* distributed throughout North America, some subspecies is to be found in almost every locality. Deserts, low mountain regions, prairies and eastern farmlands all have their characteristic forms. Among them there is a great deal of variation in color and some in size, but all conform to the general description of a small rabbit of brown to buff-gray, with no startling difference between length of front and hind legs and with the distinctive white powder puff tail from which the species takes it popular name.

The cottontails are ordinarily residents in areas where bushy or prickly vegetation offers a protective shelter underneath which they may lie during the heat of the day. From these "forms" they emerge at dusk to begin nocturnal feeding on such tender, green stuff as is to be had. In arid regions of scant vegetation and shelter, the cottontails are often found living and nesting in holes abandoned by previous tenants, or in self-dug burrows.

Several litters of from two to six helpless and unfurred young are produced each season. These youngsters often have as their birthplace nothing more than a depression in the ground, lined with the fur of their mother and concealed from detection by a skillful camouflage of dried grass. After they have begun to hop about in the immediate vicinity of the nest, they often fall into human hands. The captor may assume that their care will be simple. This is not altogether true. Even though beginning to nibble at green stuff—grasses, alfalfa, clover and the like—the young are usually not yet free from the need of mother's milk. The substitute for this natural supply is cow's milk or diluted condensed milk fed from a spoon or by means of an improvised eyedropper nipple and bottle.

Shock, sudden noises and excitement are particularly to be avoided with cottontails, young and old. Unnecessary visits to their quarters in the early days of their captivity are unwise. They are very nervous and any sort of scare sends them into a state of collapse; or they dash wildly and frantically at boards or netting in an attempt to escape, and often fatally injure them-

L. W. Walker, Arizona-Sonora Desert Museum

JACK RABBIT

U.S. Fish and Wildlife Service

COTTONTAIL

RED FOX

selves. Dogs, on this account, are anathema in the vicinity of the cottontail cage and should not be allowed to come near.

The cottontail cage should be large—at least six feet by three wide by three high—and *must* be provided with a dry, roomy, closed compartment, filled with clean straw, in which the inmates may feel safe and hidden.

Extremely timorous and gentle to a fault with humankind, cottontails seldom agree among themselves and, unless watched and given plenty of room, serious fights take place. A large outdoor pen helps to solve this problem. The wire need not be high but must be sunk into the ground a couple of feet at least to prevent escape by burrowing. A dry, covered trench or buried half-barrel lined with straw makes a satisfactory hiding place. Low-growing shrubbery is valuable for shade within the pen although, if the leaves and bark be tasty, it may not long retain its freshness.

For adult cottontails a good diet is a feeding of hay and whole oats twice daily, with an additional ration of green and root vegetables at least once a week. Cod liver oil in small quantities, mixed in rolled oats or cornmeal, should be given weekly. Clean water should be available at all times but especially if the feeding of greens is scant.

THE JACK RABBIT

Because of a host of enemies possessed of unremitting hunger and vigilance, the jack rabbits of the western plains and valleys have developed an extraordinary cunning and an acute sense of the presence of danger. The name "jack rabbit" itself refers to the outward sign of this acuteness and awareness: the long, upstanding, jackass-like ears.

Belonging to several species with many regional subspecies, the jack rabbits of the genus Lepus have at times in the past become so numerous in various sections, California especially, that organized drives resulting in the killing of thousands were necessary for the salvation of grain and other crops. The sudden and excessive increases in the number of jacks of a region have been in part due to the destruction (by traps and poison) of animals such as the coyote, whose normal prey is the jack rabbit. If not checked by human or animal means, periods of great population among the jack rabbits inevitably result in a scarcity of food and a resultant weakening of the stock. At this point disease and epidemic step in and by wholesale devastation restore the group to a more normal relationship with the environment.

Unlike the cottontails, jack rabbits have hind legs entirely disproportionate to the front in length. Like those of the kangaroo, the hind legs are for jumping, and the high, flying leaps of a jack escaping from danger demonstrate their value.

In captivity, jack rabbits feed and behave much like the cottontail. Like them, they must be protected from shock and excitement until well-conditioned to the new environment and human society.

Sick rabbits should not be kept but quietly done away with and the carcasses burned. Snuffling and sneezing are the beginning symptoms of a serious rabbit disease which often assumes epidemic proportions. The activity of this same bacillus sometimes produces large and disfiguring abscesses, contagious and difficult to cure. If not disposed of, the affected animals should be kept in a rigid quarantine.

THE OPOSSUM

The opossum has little to recommend it as a pet. It is somewhat surly in temper, inclined to dirtiness in personal habits, and decidedly unprepossessing in appearance. As a curiosity, however, a specimen belongs in every comprehensive animal collection. The opossum is our only North American representative of that group of animals to which the Australian kangaroo belongs: the marsupials. The striking characteristic which sets the marsupials apart from the rest of the animal world is the female's possession of an abdominal pouch in which the young, born at a very small and immature stage, complete their development. Squirming, naked little bits of flesh, the baby opossum attach themselves to the teats within the pouch and there, almost as secure and cut off from contact with the world as if unborn, grow into recognizable animals. They continue to occupy the pouch intermittently—sometimes as many as a dozen of them—even after they have been partially graduated to a more advanced diet than milk. Eventually, the pouch becomes too small to hold them; the mother becomes weary of their weight and refuses to admit them to the nursery. Until they are able to take care of themselves, the mother carries them about hanging to her back by feet and tail. This latter possession of the opossum, a prehensile tail used for grasping, is an unusual appendage among American animals and a very convenient one.

The opossum is largely a creature of the night and sallies forth to forage after dark from its den in a hollow tree or a hole. Generally, it is fond of lowland areas, along stream beds especially, where low waters offer good hunting and fishing.

To escape from danger, the opossum depends mostly upon his climbing ability, but, when cornered, gives an exhibition of the well-known trick "playing 'possum." This imitation of death is most convincing and complete. However, there is reason to believe that the opossum, instead of consciously fooling, actually is so overcome by fear as to pass into this surprising and protective coma.

Opossum, young or old, are easily maintained in captivity, and except for the difficulty of keeping them clean, present no great problem. They do little if any self-grooming, and in a small cage, would rather go to sleep in a dish of partly consumed food than anywhere else. In an outdoor pen, they are cleaner and, after they have had a chance to become accustomed to the presence of people, lose to some extent their nocturnal habits and put in day appearances. Their food may be almost anything available: bread, vegetables, meat, fish, or, in the case of the young, simply bread and milk. Feed once daily in the late afternoon.

The range of the Virginia opossum, *Diadelphis marsupialis,* was originally through nearly all the wooded parts of the eastern United States. Within the last few decades they have been introduced along the west coast in several localities. In about 1908, I saw a half-dozen opossum, which had been shipped to a local saloonkeeper from his old home in Missouri, liberated along the banks of the Stanislaus River in California. By 1929, they had become numerous enough to be found in vineyards a couple of miles from the river bottoms and were not infrequently killed by cars on the highways.

FOXES

The memory of a red vixen tethered alongside a ramshackle service station in South Carolina really prompts the writing of this section. Just out of range of her rope, wary and suspicious chickens picked at trash. A dozen feet away, a rusty old Ford coupé, stripped of everything of value, was the home of a filthy squad of monkeys with nothing better to do than scratch, squall and throw bits of rotten upholstery at one another. A tired rattlesnake sprawled its skinny and sick length across the bottom of a makeshift cage atop an empty gasoline drum. High above everything, in letters three feet high, a sign proclaimed that here was a zoo. In the midst of it all sat the fox, calm and disdainful, ignoring alike the insults and smell of the monkeys, the temptation of the chickens and the curiosity of customers.

We asked her history and were told by the proprietor that he had dug her from a den as a cub. Introductions followed and, these formalities over, we found her as friendly as any dog of reserved disposition. She was obviously attached to her owner, much more so than he to her, for, noting our interest, he promptly offered to sell his pet. Her coat was poor and she obviously would have been in better condition had one of those very wary chickens occasionally fallen to her lot. According to her owner, her diet consisted mostly of "odds and ends."

In nature, the food of this species of fox, *Vulpes fulvus*, is made up of "odds and ends" of an altogether different sort from the poor garbage the owner of this animal meant. Insects, small rodents of all sorts, ground-dwelling birds, domestic fowl when available and other animal foods, together with casual portions of fruits and berries, make up a rich and healthy fare.

Had this little vixen been fed her proper due in half-cooked meat, dog biscuit, milk and an occasional bit of fruit, instead of the meager and watery kitchen waste we saw in her pan, she undoubtedly would have been a much larger and handsomer animal.

The red fox is fairly abundant in parts of the United States, more so in the East than in the West. Contrary to being the economic hazard the farmers complain it is, the species is valuable for its help in rodent control. In spite of the fabulous prices its skins bring, the famous silver fox is nothing more or less than a black phase of the red fox. The breeding of this and other color phases in captivity has become an industry. Special publications are devoted exclusively to problems of feeding and housing these and other fur-bearers. A number of U. S. Department of Agriculture bulletins dealing with the same problems are available on request.

A number of species and several subspecies of the gray foxes of the genus Urocyon range through most of the United States. Their food habits are much like those of the red fox and, in spite of their value in rodent control, they bear the same burden of farmers' ill-will because of an occasional stolen hen. Unlike the red fox, the gray foxes are climbers and are often seen spy-

U.S. Fish and Wildlife Service, photo by Allen M. Pearson

GRAY FOX, APPROXIMATELY HALF GROWN

L. W. Walker, Arizona-Sonora Desert Museum

COYOTE

WILDCAT

C. M. Bogert

U.S. Fish and Wildlife Service, photo by Frank M. Blake

FEMALE OPOSSUM WITH YOUNG

ing out the landscape from a vantage point in the crotch of a tree.

THE COYOTE

The name of the coyote, the little prairie wolf, *Canis latrans,* is derived from the ancient Aztec "coyotl" and in a number of regional pronunciations is used throughout our country to refer to this close relative of the domestic dog. Early travelers and explorers of a timorous nature found its nocturnal howls hair-raising and frightening and attributed the noise that so disturbed them to some mighty bloodthirsty monster. It is true that one coyote can sound like a dozen and several can fill the night with noise, but, to me at least, a chorus is melodious and, even in the very early hours, a pleasant addition to the desert night.

The food of the coyote, although consisting mostly of smaller game such as rabbits, squirrels, mice and rats, may include deer and even domestic stock. This larger game, however, is usually crippled or diseased and the coyote, in many cases, is eliminating a possible source of infection of healthy animals. In rangelands where the coyotes were completely exterminated, grass- and seed-destroying rodents so increased in a few years as to leave no food for sheep, cattle or the deer. The result—a desolation which eventually caught the overpopulation of rabbits and other rodents in its trap.

Coyotes, especially in their younger years, make very good pets and seem to have some unexpected qualities and abilities. They can cooperate with one another in the carrying out of hunting plans and, when not engaged in more serious doings, singly or in groups indulge in spirited play. Although there is some disagreement in the matter, my own school of thought holds that coyotes have a definite sense of humor and can appreciate a joke, especially when it is on someone else.

A very responsive female of my acquaintance likes to play a game of trying to gain possession of some such odorous object as a cigarette pack held in the teaser's hand. The pack is snapped at with a great display of wrinkled nose, flattened ears, crouching and even, on occasion, a most theatrical sneeze. If she manages to grab the pack—for economy's sake I never play with a full pack—she tosses it into the air a few times, then quickly and triumphantly tears it to shreds. Then, tongue hanging out, head tilted to one side, she sits back and looks at the loser with a most quizzical gaze. She knows that we know she has put one over on us and, all in good fun, thoroughly enjoys the trick.

THE WILDCAT

The wildcat, or, as it is known because of its short tail, the bobcat, *Lynx rufus*, ranges throughout most of the wooded United States in a number of subspecific forms varying in size, color and choice of habitat. Although among our most widely distributed animals, wildcats are seldom seen because of nocturnal habits, extreme cunning and protective coloration. The commonest indication of their presence in a locality is their tracks, much like those of a cat, but larger, in the dust of roads and trails.

Slightly tufted ears, short tail, ruffed jaws and lightly spotted coat all show the relationship of the wildcat to the larger Canada lynx.

Like all the cats in the wild, bobcats are carnivorous. All sorts of small game and rodents help make up their fare. Barnyard fowl occasionally suffer from their depredations. In captivity meat and fish are their main diet. However, chicken heads and the whole carcasses of rabbits, rats and the like provide a welcome change and a necessary one. Meat alone as a ration lacks the completeness and balance to be found in the whole carcass.

Even very young wildcats spit, snarl, scratch and bite when captured. Older ones, through selling their lives and liberty dearly, have given rise to all sorts of legends concerning their scrapping ability.

As a captive, the wildcat is often seen in roadside zoos, confined and cramped in small quarters. Under these conditions, with no hiding place, pestered by people from all four sides, its temper is seldom mild and visitors are met with a baleful glare and a growl. In a larger cage, with room to exercise and with a box or barrel to retreat to, the bobcat's temper is less surly.

Although baby bobcats can be gentled and stay relatively tame, they should not be trusted too far when grown. Small and accidental factors often cause them to revert to a primitive pattern of behavior based upon the rule of tooth and claw.

Small Mammal Problems

Young animals that have never known the freedom of the woods or the dangers of being too trusting are by far the best additions to an animal collection if docility and gentleness are prime considerations. Adopting young, however, is a precarious business and once that calls for nice judgment. If they are too young and have not as yet added to their ration of mother's milk a reasonably large part of the adult diet, the case is almost hopeless unless the adopter is very patient and sufficiently interested to break his sleep of nights and give the children their bottle. Even with such attention and care, survival of animals taken too young is most doubtful. The kidnaping should be timed to coincide with the youngsters' natural urge to get out and see what the world looks like. The difficulty of locating the nests of desired species at just the right time to get young sufficiently advanced to manage a change of diet makes nest hunting a haphazard means of collecting.

Preliminary investigations of nests should be made cautiously and always when the mother is away. Squirrels, especially, have been known to remove the young to a new nest after having been upset by a discovery of their home and too much curiosity on the part of the discoverer. Too, unless the collector is reasonably certain that the lady of the house is abroad before he ventures to reach down into the darkness of a tree hole nest, his finger may become a bloody and scarred Exhibit A in a demonstration of the squirrel's protective care of the young.

TRAPPING

For small game, mice especially, a simple trap, efficient and easy to make, may be rigged out of an ordinary spring mousetrap, a piece of wire mesh and a tin can. Traps similar in principle may be purchased in any hardware store, but these are less satisfactory than the homemade article; the trigger is set too far forward in them and instead of being caught in the trap unharmed, the animals are caught in the door and killed, or at least suffer pinched tails. Small animals in traps are particularly

49

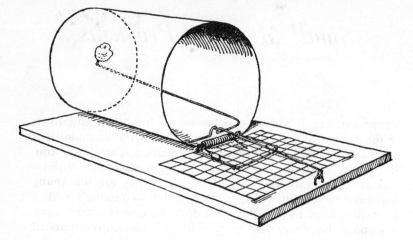

TIN CAN MOUSETRAP

susceptible to the effects of exposure, and to protect them the trap should contain a wad of waste, a bunch of grass or other soft bedding to roll up in.

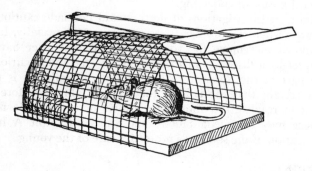

MOUSETRAP

Box traps are most useful for larger animals and are especially to be recommended for cats. The domestic cat wears its domesticity with a difference. In addition to protection, a seat by the fire on cold evenings and regular portions of canned salmon, the cat clings to its wild rights and regards as legitimate prey any small creature unfortunate enough to come within reach of its

CAT TRAP

claws. Any attempt to make an area attractive to small game begins with the elimination of stray cats and the control of those with owners and homes.

Several types of box trap—all fairly efficient and simple—may easily be made by anyone able to drive a nail and saw boards square. Their great drawback is a heaviness which makes them impractical far from home, and a conspicuousness which renders them less useful for shy and wary wild creatures. Seemingly good sets will often prove disappointing until the animals have become somewhat accustomed to this threatening rig plumped down in their haunts. If a box trap is set for squirrels, rats, porcupine or other large rodents, it will be necessary to line it inside with either sheet metal or heavy wire cloth to prevent their gnawing out. The Allcock Manufacturing Company, Ossining, New York, makers of the Havaheart animal traps used by many naturalists, will send free on request their Trapping Booklet.

The setting of traps for specific animals requires some knowledge of the habits of the creatures desired, their foods and prejudices. Small animals in the wild are almost invariably shy, fearful of strange objects and smells. Their nocturnal traveling is cautious and, for protection from flying and creeping enemies, they stick closely to the concealment of logs, stumps and other cover. Traps set in the lee of such cover are in favorable locations. In open ground, meadows, swamplands and the like, the runways and customary paths of small game are obvious trap sites.

A little experience and catch-as-catch-can trapping soon gives the collector an idea of the animals to be found in his locality and their ways.

Baits for traps may be almost anything in the way of food attractive to the animal desired and *not* found in any abundance in the locality. Peanut butter is a good general mouse and rat bait, but many other edibles such as bacon rind, pieces of apple, rolled oats, bread and cornmeal have their value.

The careful trapper visits his traps daily and if the set has been unsuccessful, cleans and rebaits them. Ants and other insects find food and carry it off and night dews, by a slight corrosion of the metal parts, tend to make a long-standing set ineffective.

Pit trapping is an ancient device of man and, although the most primitive means of taking animals alive, is still a practical method for capturing large and dangerous game. While it is reasonably certain that few of us will ever be involved in a tiger or elephant hunt, it does not follow that pit traps are out of our line. A small open hole in the ground is to a mouse, a mole or a shrew as a deep, masked pit is to the tiger or elephant.

Pits should be straight-sided, perhaps eighteen inches deep and twelve or fourteen inches wide. They should be dug across the runways of mice or in the anticipated paths of the shrews or other game desired. A rainstorm of a few minutes' duration is enough to drown any small pit-caught creature, and so a piece of wood for a float should always be left in the bottom. Death from exposure, or from starvation, even if the pit is layered with grass, is likely to occur if the pit is not visited regularly. The trap is not selective and a creature of no particular value or interest may slaughter or trample a good specimen of a desired species. Naturally, the attempts of small animals to escape are incessant. Mouse philosophy says "try, try again" and the wise trapper visits his trenches often.

When a pit is no longer needed the trapper should feel it his duty to fill it. Unattended, it is not only a continual menace to the small woods creatures, but larger animals may seriously injure themselves by stepping into it.

FEEDING

One of the most important factors in the keeping of animals is conscientious routine care and regular feeding. Food, in most cases, should approach the animal's natural diet, not only in chemical and vitamin content, but also in texture and toughness. The incisor teeth of gnawing animals continue to grow throughout life and, by wearing against one another and against food materials, are kept at just the right length and at just the right sharpness. If fed exclusively on soft foods which offer no chewing resistance, these teeth sometimes become seriously overgrown and distorted. Rodents occasionally injure the jaw or knock the teeth out of line so that they fail to meet and wear away properly. In these cases the incisors go wild; their excessive growth makes feeding more and more difficult and eventually the rodent perishes from starvation.

The rate of growth of small animals, rodents especially, is very fast as compared to that of the larger creatures. Feedings must be regular and frequent to satisfy heavy growing needs. For very young squirrels, rats, mice and the like, sweetened condensed milk, diluted with warm water according to instructions on the can, is good. Cow's milk is seldom satisfactory except for babies that already have a good start and are beginning to experiment with adult diet. A little lime water every day does no harm. A drop of cod liver oil every other day tends to prevent rickets.

Feeding from a spoon is messy. An eye dropper, with a bit of fine rubber tubing over the end to obviate any danger of the growing teeth shattering the glass tip, may be used for very small creatures. For animals large enough, a nipple made by piercing the bulb of an eye dropper and used on a small bottle simplifies matters. To avoid soiling the fur with milk, wrap the infant in a cloth while feeding.

Laboratory rats and mice have of late years become an increasingly important element in medical and biological research. Several firms, formerly manufacturing only dog and cat foods, now have on the market rations especially formulated for the feeding of these animals. These are inexpensive and offer a convenient food, properly balanced and with an adequate vitamin content. In texture and composition, they are much like miniature dog biscuit. If they are not easily obtainable, ordinary dog biscuit broken in small pieces, with the addition of lettuce or other greens, makes a good ration, not only for the common rats and mice but for the wild species as well.

Calf meal is a balanced ration intended to take the place of whole milk in the feeding of very young calves. In practice, it proves a most satisfactory and economical rat and mouse food but, being soft and powdery, gives the teeth no workout. To remedy this defect in the diet, add cooked bones from the table.

A ration of this type should be fed daily in quantities sufficient to insure food being present at all times. Use shallow dishes for containers and see that clean water, preferably in a drop bottle, is always available. No lettuce or greens is needed with such a balanced ration but may be fed twice weekly to relieve monotony. Small creatures that nurse their young will fare better if they are fed an inch cube of beef once a week.

LIVING QUARTERS

The problem of living quarters for his charges is one that each keeper will, in all likelihood, have to meet and solve for himself. Babies should be kept in a warm, dry box, with plenty of cotton batting, dry grass or the like for bedding. This should be changed as it soils.

Several types of rat and mouse cages are made for laboratory use by biological supply houses. Pet shops and animal stores have ready-made cages suitable for squirrels and larger small mammals. For the most part, however, it is safe to assume that the budget will not permit the purchase of these rather expensive and not entirely satisfactory setups. The laboratory cages, while excellent from the point of view of cleanliness and practicality, do not afford a good view of the inmate's activity; the heavy wire mesh obstructs vision and renders photographic work almost impossible. Glass-sided aquaria offer no insurmountable obstacles to the making of pictures and afford a good view, but, unfortunately, are difficult to keep clean, are poorly ventilated and, except for burrowing mammals and such special cases as the shrews, impractical.

A hybrid between the screen laboratory cage with its wire bottom, through which droppings and dirt fall into the tray beneath, and the glass container is the best answer to the problem. Shop facilities, carpenter skill and materials available determine the elaborateness of a cage, but there is no need to spend heavily to make a good one. A little ingenuity, wire, glass

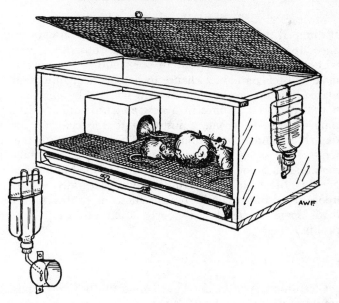

GLASS-FRONT CAGE WITH WATER BOTTLE

and wood are about all that are needed to turn a stout box into a perfectly satisfactory cage.

Bedding should be of coarse shavings, excelsior or shredded paper. Cotton is satisfactory in some cases, in others inadvisable; excellent as bedding, it has no good effect when taken into the digestive tract.

To lessen the odor always present around mouse colonies, the cage tray should be well covered with sawdust. A specially prepared, pine-scented compound for cage use is on the market. Rats are less heavy in odor than mice and, given adequate room and cleaning, are not so likely to become offensive.

The most necessary cage furnishing other than food and water is a hiding place. This may be nothing more than a small box with a hole in it just large enough for the animal to enter.

Rats and mice manage to get exercise in captivity without mechanical aid. However, they appreciate and use an exercise wheel. For squirrels, an exercise wheel is almost essential, unless the cage is large enough to hold branches and perches far enough apart for jumping and racing.

Most species of wild mice are more or less solitary in habit. If dumped together for breeding with no preliminary acquaintanceship the results are sometimes disastrous. Kept in separate wire-screened cages adjacent to one another for a week or so, male and female have a chance to become acquainted and when placed together, ordinarily take up family life amicably. Or, if both are placed at once in a new and unfamiliar cage, by the time they have become accustomed to the changed surroundings, their murderous impulses have worn off. With the white-footed and other small wild mice, mating should take place soon after a brood of young have been weaned and taken away. Fertility of small rodents in captivity naturally is dependent upon many other factors besides those of proper housing and food. Some species breed poorly, if at all, while others are prolific.

CARE

Cleanliness, proper food, an opportunity for exercise, and ample sunshine are about all the animal keeper can do for the welfare of his wards. Any attempt to treat their diseases is almost hopeless, for by the time the ailment is obvious, the animal is usually so severely affected that a cure is impossible. Our best hope is by proper care to make our charges less susceptible to the common pulmonary and intestinal diseases which threaten animals of lowered resistance.

Cleanliness includes freedom from external parasites. As a routine procedure, it might be well to give all new rodents a disinfection and keep them away from the rest of the colony until it is certain that they do not harbor fleas or scabies mites. These latter tiny, microscopic lice, are a serious matter. In 1920-21, a scabies epidemic among the western gray squirrels almost exterminated them and it was feared for a time that the species would not be able to regain its foothold. An animal with scabies loses hair and is continually scratching at red and inflamed areas. The ears become affected and scabbed. The parasites are hard to eradicate but it can be done. Cages should be washed with hot soap and water and scalded. If possible, treat with gasoline. The animals should be treated with sulphur ointment as prepared for human use. Several applications may be necessary.

Fleas, although causing much discomfort to animals, have no such capacity for damage as the scabies mite. They may be eradicated quickly and certainly by the use of any good rotenone-pyrethrum powder not contaminated with DDT.

In the wild, rodents are often found infected with the large larvae of a bot-fly. In these cases, an operation beyond the skill of any except the trained person is necessary to remove the parasite. For the ultimate good of the unwilling host and to avoid adding a weakened individual to the animal collection, it is best to do away with the unfortunate or turn him loose to work out his own salvation.

When keeping wild animals, the welfare of the owner and his friends must be considered in addition to the animal's well-being. In the case of raccoons, skunks, foxes, coyotes, bobcats, and even perhaps the squirrels, a veterinary should be consulted on the advisability of vaccination for rabies, distemper, hepatitis, and other ailments. Diseases formerly confined largely to domestic animals have seeped into the wildlife population and present a serious problem in wildlife maintenance.

Several successive epidemics of rabies have swept our country in our own time, affecting many species of wild and domestic animals and providing a possible reservoir of human danger. If, in the woods, along the roads, in public parks or your own backyard, a strange wild animal appears—or even a domestic one—be cautious and don't risk a bite. In many cases of rabies, in skunks and foxes particularly, the animal may seem so tame and placid as to invite petting. DON'T. Actually, it may be "dopey," certain to bite if touched and as dangerous as a creature in the frenzied or "mad" stage.

Skunks are susceptible to rabies and a veterinarian should inoculate one of these animals ONLY with a dead vaccine as the attenuated live vaccine often used may possibly cause trouble.

Bats, although performing a useful function in the control of insects, are not very prepossessing at best, difficult to care for, prone to bite and notorious carriers of rabies. Rabid bats from all parts of the country have been reported and they had better be left to their nocturnal mosquito and bug catching. Never try to pick up a bat. The one you can catch is likely to be sick and exactly the one you should not touch.

SHIPPING

The collector of small mammals may wish to enrich his collection by a trade with someone in another locality. The

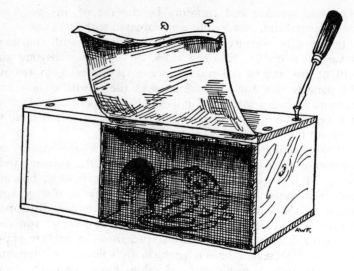

SHIPPING CAGE

shipping of live small animals, unless carefully done, can be a disappointing business. Shipping cages should be of two compartments: a feeding chamber and a smaller and dark section where the animal can hide and sleep. A shipping cage of wood should be lined with wire mesh if the animal is inclined to gnaw. A wire mesh front, covered over with a flap of burlap or with strips of wood, permits some examination of the contents by curious carrier company employees, without it being necessary or excusable for them to open the box. The water container in such a cage, if one is needed, may be filled through the mesh by means of a funnel. Water dishes, if used, should be firmly wired into a corner.

For mice, rats, squirrels, kangaroo rats and the like, a water dish is not necessary and had best be dispensed with. Small pieces of apple, carrot or other succulent vegetable previously found acceptable to the animal will furnish adequate moisture. Whole apples or other vegetables, while remaining in better condition for a longer time, tend to roll about the cage. Too large a ration only dirties the inmate and the cage. Be sparing of what is put in.

Traveling is tiresome for all of us, and in cramped and crowded quarters creatures normally amicable and friendly

sometimes turn against one another. The result is that ship-
ments sometimes arrive with but a sole survivor and that
survivor in no condition to live long. For this reason, it is best
to give each small creature his own section of the shipping
cage, his own food supply and his own little hiding can or
chamber.

Birds

IN MOST cases, being ornamental is one thing; being useful quite another. One form of our wild life, the birds, combines usefulness and beauty in proportions which have leagued in their behalf widely diverse elements. Individuals and organizations early interested themselves in the conservation of bird life for their song, grace of body and flight, and harmony of color. This movement was preceded by a recognition of the birds as valuable "farmer's friends," indispensable in the destruction and control of weed and insect pests.

The result of the combined forces of practicality and beauty is that we have a strict and comprehensive code of State and Federal conservation laws which protects our native birds from many hazards at the hands of unthinking and careless citizenry. These laws effectively prohibit the killing, trapping or keeping in captivity of most of our feathered residents.

In spite of this, a large collection of live native birds is by no means impossible to an individual interested enough to keep them and to provide for a part of their needs. They may not be kept in cages; the whole outdoors must be the aviary; but given a little encouragement, nesting facilities, food and protection, numbers of them are willing to attach themselves to any person offering these gestures of friendship. Any suitable premises made safe and attractive to them becomes, in a very short time, the scene of the unfolding of countless small dramas of feathered life.

Spring brings with it large flocks of birds returning from their winter feeding grounds. The males of most species return first and begin to explore the countryside in search of suitable territories and, after the females arrive, of mates. As a rule, it is not difficult for them to find the mate—song and an adequate display of masculine beauty take care of this—but a home is a different matter. In this day and age, the activity of the tree surgeon in every wood lot and an overzealous thinning out of dense shrubbery have eliminated many natural nest sites. This is a problem which we may solve and turn

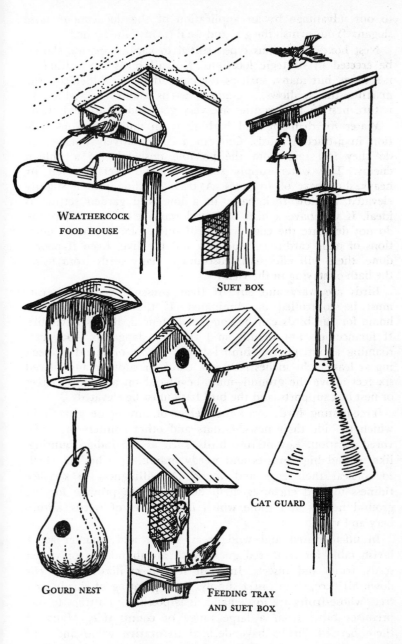

WEATHERCOCK
FOOD HOUSE

SUET BOX

CAT GUARD

GOURD NEST

FEEDING TRAY
AND SUET BOX

to our advantage by an application of the department store slogan: "You furnish the girl and we'll furnish the home."

Nest boxes of various types, suited to various species, should be erected in strategic locations. Not all birds are willing to use them, but many, such as the chickadees, the bluebirds, the martins, the swallows, the house wrens and others which in nature nest in tree cavities are very glad indeed to find them.

Water for drinking and bathing is an important consideration in attracting birds. Only in a comparatively open space do they feel safe from the possible onslaught of a hidden enemy. The water supply should not be overshadowed by heavy shrubbery or by trees. A trickle falling into a slightly elevated, shallow rock basin in a low wild garden setting is ideal. If you have a bird bath and want it to be used, please do not decorate the edge with brilliantly colored plastic imitations of male cardinals, blue jays and the like. Even if poorly done, these will effectually discourage many birds from using the bath or staying in the vicinity.

Birds are wary and shy. If their presence is desired, cats must be controlled or eliminated. If the area intended as home for the birds can be completely fenced, this is advisable. If finances do not permit, and there is danger of stray cats roaming about, guards—funnel-shaped pieces of metal projecting at least eight inches from the trunk or support and placed six feet above the ground—must be placed on all nesting trees or nest box supports, and the bird bath must be elevated.

Tree-nesting birds avail themselves of any dense growth in which to hide their nests. Cedars and other conifers are a favored location. Low-nesting birds, such as the indigo bunting, like to find high grasses and weedy bramble patches in which to conceal their eggs and young. A willingness to sacrifice tidiness to bird encouragement does much to provide for the ground nesters and those which like tangles of weeds, shrubbery and vines.

In uncultivated and wild areas, spring and summer set a lavish table for feathered guests with an abundance of varied seeds, fruits and insects. In most regions civilization has cut down all except the insect supply. By a planting of shrubs and trees whose fruits are in demand, it is possible to attract to our premises birds from a large range of countryside. Many of these berried shrubs have decided decorative value and add their own beauty of leaf, flower and fruit to our grounds. The following trees and other plants hold fruit well into the winter:

Bayberry, *Myrica caroliniensis*
Birches, *Betula sp.*
Bittersweet, *Celastrus scandens*
Black alder, *Ilex verticillata*
Box-elder, *Acer negundo*
Bull thorn, *Smilax rotundifolia*
Coral berry, *Symphoricarpos orbiculatus*
Corkbark, Chinese, *Phellodendron sp.*
Crab, Flowering, *Pyrus floribunda*
Grape, *Vitis labrusca, V. vulpina*
Hackberry, *Celtis occidentalis*
Hawthorn, *Crataegus crus-galli, C. intricata, etc.*
Holly, American, *Ilex opaca*
Hornbeam, *Ostrya virginiana*
Inkberry, *Ilex glabra*
Mountain Ash, *Sorbus americana*
Mulberry, *Morus rubra, M. alba*
Roses, *Rosa spinosissima, R. rugosa, R. blanda,*
    R. setigera, etc.
Snowberry, *Symphoricarpos alba*
Sumac, *Rhus copallina, R. glabra, R. typhina,*
    R. toxicodendron, R. vernix
Virginia creeper, *Parthenocissus quinquefolia*
Winged euonymus, *Euonymus alatus*

Winter feeding stations may be of several sorts. A three-foot square platform on a stake high above the reach of cats is the simplest. To keep the food from rolling or being blown off, it should have a low rim around it. A further improvement is a shelter over it to keep snow from covering the provisions and to protect the feeders from rain. More elaborate stands to suit the taste and the finances of the builder can be bought ready-made or built. Wide window shelves and stations close to the house have the advantage of permitting observation of the guests without disturbing them and without the observer being subjected unnecessarily to cold.

A particularly ingenious and practical arrangement for winter feeding has been devised and used with success. In this case, the feeding stand, called the "terravium," covered by a glass case, rests on a window sill and is open to the outside to

"TERRAVIUM"

admit the birds. Instead of being a barren shelf the "terravium" is cleverly landscaped. Lichens, stones, gnarled roots and branches, hardy ferns and mosses make it a small bit of winter woods. Attached to the branches are bits of suet fastened with rubber bands. Seeds, especially those of the sunflower, are scattered over the feeding area. A flat pottery dish of water, frequently warmed, is a welcome relief from the icy puddles. Even such shy birds as the hairy and downy woodpeckers visit the "terravium" frequently, while the tufted titmouse, the chickadee and the nuthatch are constant visitors.

Early fall is the time for setting out the feeding stations for the winter birds. Returning from the north, many species are in search of a satisfactory wintering place and unless their appetites and interest are engaged, may pass on. At this time, being in a strange locality, they are apt to be shy until they become acquainted with the premises and the proprietor. For their peace of mind and comfort, feeding tables should be set in sheltered locations, with easy escape to the protection of shrubbery available. If no shrubs of sufficient size are close to the feeding table site, piles of evergreen boughs or a shelter made of poles draped with vines and cornstalks form a good hiding place for the shyer species as well as for the song sparrows, white throats and juncos. Shelters should be open in several places in order that birds may escape from any marauding cat.

Weed seeds are a natural food of many of the winter birds. Cut green, the seeds ripen but do not fall. Such brush with its pendant food is good shelter material. Inside the shelter, coarse sand or small gravel should be provided. Birds use such grit in their digestive processes to grind up food and often, when everything is covered with snow, are hard pressed for this necessity. A little scooped-out place filled with soft, dry loam provides a dust bath.

The covered feeding table should be close to the shelter. For better protection from cats, it may hang from a tree or be supported by a post with a cat guard. Many designs are on the market; they can also be made with but little time and trouble. An interesting innovation in feeding trays, sponsored by the Audubon Society, is the weather-vane type which has but one side open and swings so as to present always the walled side to windward.

To coax shy birds into coming closer to the house for observation and more intimate acquaintance, a trolley feeder is the right type. This is simply a small roofed shelter hung on a pulleyed wire between the window and a tree or post. After having been drawn in for replenishing, the tray each day may be left a little nearer the window. Eventually, the birds sit on the sill, and finally, in the host's hands if everything goes well.

The matter of provisioning the feeding trays and tables is simple. Birds, unlike some of our smaller creatures, are not difficult to feed. For those which, like the woodpeckers, feed largely on insect larvae, nothing is better or more tasty in cold weather than suet. It is so tasty that in order to protect the

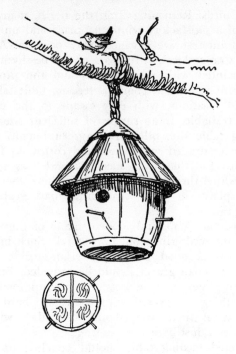

HANGING BIRDHOUSE

CROSS-SECTION OF SAME HOUSE, SHOWING FOUR SEPARATE NESTS

supply from those pirates, the blue jay and the red squirrel, a dispenser is necessary. There are many devices for this purpose. The cheapest and most easily procured is an ordinary wire soap holder. Any sort of coarse wire mesh container which will prevent the whole lump of fat being stolen will serve the purpose.

Other birds besides the woodpeckers are steady customers of the suet supply. The nuthatches and creepers enjoy both suet and chopped meat scraps. Even the song sparrow may be seen, clinging to the wire, pecking at the fat inside.

Suet alone is not enough. Many birds, such as the sparrows, are seed-eating and for their provision as well as a better balanced ration for the chickadees, nuthatches, evening grosbeaks, crossbills and the like, an appropriate menu must be supplied. Bread crumbs, chick feed and crushed dog biscuit are good. Seeds, particularly those of the sunflower, hemp and millet, are most welcome. Nuts and cracked corn are choice tidbits.

A novel feeding device, easily made, is a pine cone rolled in half-melted fat into which, after the fat has congealed, nuts and seeds are pressed firmly. These cones may be hung from trees or from supports.

Small Christmas trees, after they have outlived their short holiday splendor, may be utilized in a very practical fashion for feeding in the manner recommended by the Audubon Society. Set securely in the ground or on a standard, the tree is covered with a mixture of fat and bird food. The formula advised is an adaptation of the well-known Von Berlepsch mixture:

Bread, dried and ground	5	oz.
Meat, dried and ground	3	oz.
Hemp seed	5	oz.
Millet	3	oz.
Ant eggs	2	oz.
Sunflower seed	3	oz.
Dried berries	1½	oz.

Add to this mixture one and one-half times as much melted suet or fat. After stirring the mixture thoroughly, pour over the branches of the dead tree.

The permanent bird residents of the eastern states are chickadees, woodpeckers, song sparrows, the white-bellied nuthatch and gold finches; sometimes cedar waxwings, robins and flickers are also to be seen. Among birds which winter with us but go farther north for their nesting are juncos, tree sparrows, siskins, kinglets, myrtle warblers and the red-bellied nuthatch. Some seasons we have with us the evening and pine grosbeaks, the redpolls and the crossbills. Some birds which ordinarily go south for the winter, finding weather mild and food plentiful, stay over.

By providing a safe refuge and adequate food supply the student can increase the normal bird life of a locality several fold and open up for his delectation a rich store of firsthand natural history information. And, if he has a clear eye and open mind, he may find that although St. Francis preached to the birds, the birds, in turn, have sermons for us.

The laws protecting birds are almost fully inclusive. Formerly outside the pale was a large group condemned to be hunted because of some real or fancied conflict with man's interests. This group of rogue birds included, in practically all states, a number of species of hawks and owls, the crow, the magpie, and the blue jay. Through the years, a little common sense has crept into the respective attitudes of the public and the law and now the crow is left as the sole official renegade and unprotected bird.

The rascal birds are only rascals in a man's world. In their own, they are hard-working, respectable citizenry, trying to get along in the fashion for which nature fitted them. With every chance shotgun a hazard to life in the wild, we may be doing the bird a favor by taking him out of circulation.

Legislation varies and it would be well before capturing any species to check its status in your locality and, if needed, secure special permits from your state department of conservation.

HAWKS

The hawks of the genus Falco are of a noble race and were for centuries the prized hunting companions of kings. The sport still survives in the Orient and in Northern Africa and there has been a limited movement to popularize it in the United States. The whole history of falconry is evidence that the birds of this group are extraordinarily intelligent, teachable and courageous.

The falcons are bold in nature and carry assurance and courage into captivity. The larger American falcons, the gyrfalcon, *Falco rusticolus,* and the peregrine falcon, *Falco peregrinus,* are rare, increasing in demand among falconers and not likely to fall into our hands.

The third and smallest of the American falcons, the sparrow hawk, *Falco sparverius,* is fairly common throughout most of the United States and has the bravery and self-confidence of its larger brethren. With no great fear of man to start with, it tames readily and soon learns to accept food from its master's hand and to come at his command.

Young, taken from the nest, are naturally the easiest to gentle, but wild-caught adults—"haggards," as the falconers call them—soon lose aggressiveness and become friendly.

The favorite nest location of the sparrow hawk is in a hole or crotch cavity of a dead tree standing slightly apart from a wooded area. A pair often bring up young in the same place year after year. Attentive parents, the old birds are usually somewhere in the vicinity and put in an appearance if danger threatens their eggs or young. However, it is safe to investigate the nest without fear of an attack. Out of the three to five young usually found, the conscientious person will be satisfied with but one or two. These should not be taken until well plumaged and almost ready to leave the old homestead.

Two other species of hawk commonly found are the sharp-shinned hawk, *Accipiter striatus* and Cooper's hawk, *Accipiter cooperii*. Neither are satisfactory in captivity but remain wild, and no matter what dainty morsels are set before them, display a small peckish appetite.

THE SCREECH OWL

Certainly if we are what we seem, the owls are a wise race. Sitting solemnly at the back of a cage or on a tree limb in the dusk, an owl gives the impression of having just concluded a lengthy summing up of mankind's sins. At any moment it seems that we may expect a judicial clearing of throat and a pronouncing of the sentence.

The little screech owl, *Otus asio,* while less grave and formidable than some of his fellows, has brought down on his own head a load of legend and superstition. The call of any owl, sounding through the night, is doleful and seems to cry for an answer to all the questions to which there is no answer. The call of the screech owl adds to the query a burden of immediate woe and in its wail, eerie and weird, are compacted all the smaller fears of the dark and the strange. In simple country districts, its plaint is supposed to be a warning of coming death and misfortune. At its note, the southern Negro slides out of a warm bed and, blowing to a flame the ash-covered embers on the hearth, tosses into the fire a horseshoe or a nail to ward off the threatened danger. The French Canadian starts from sleep at the quavering sound outside his snow-banked cabin, crosses himself and murmurs a prayer. More practical, the Ozark mountaineer shivers, mutters under his breath and, barefooted, sneaks outside with the shotgun to look for the bird of ill omen that has broken his sleep. And all this over what is to the owl a simple—and probably very cheerful—song!

To justify the carrying quality of its call and its reputation for ill-doing the screech owl should be large and imposing.

Instead, it is one of the smaller species and usually may be identified by its reddish coloring and the horn-like ear tufts. In some cases, the plumage is gray rather than red.

The nest of the screech owl is made in an abandoned woodpecker hole, in the cavity of a hollow tree or even in some corner of a deserted or little-used building. The nest is not well and carefully constructed but consists mainly of such litter and junk as happen to occupy the site when the pair takes possession. Owls have a quirk to their food-handling which makes the locating of their nests less difficult than the homes of most birds. The residue of bone and hair left in the stomach after the digestive fluids have done their work is cast up as an oval pellet, instead of passing out through the intestinal canal. The ground around an owl's nest is strewn with these castings. Hawks have the same habit but to a lesser degree. The owl pellet is a shaped mass; that of the hawk is but a loose and unformed casting.

When first hatched the young are covered with soft white down and look like animated powderpuffs. Their parents are generous providers and the nest is littered with the corpses of small rodents. When the female is with her eggs or young, she often refuses to budge. If the nest is in a tree cavity or other hole, it is almost impossible to get her to leave. Parental solicitude has in some cases brought screech owls to the point of threatening attack on human beings intruding too close to their nest.

As the juvenile plumage comes in, the down quills ride the tips of the new feathers and seen against the light, surround the youngster with a curious and somewhat comical halo. At this stage, young screech owls may safely be taken and adapt themselves readily to life with human beings. They quickly become gentle and are entirely satisfactory and amusing pets.

Although able to get along with people, screech owls, even when very young, have difficulty in maintaining complete harmony among themselves and to put two in the same cage is to invite murder.

THE GREAT HORNED OWL

The great horned owl of the eastern states, *Bubo virginianus*, is hardy in captivity but, being large and intractable, is not as desirable as the less spectacular screech owl. However, cases have been reported of great horned owls which had been taken very young becoming quite tame and dependent upon human companionship.

This owl has unquestioned courage and fierce appetite. For, in addition to levying a heavy toll upon all except the larger woods creatures, it has the temerity to kill and eat skunks habitually and lives in a perpetual cloud of skunk scent. This skunk-hunting habit has in a number of instances brought human beings and great horned owls into situations of mutual surprise and shock. A white hat, seen from above in the night, apparently looks like the bobbing white stripe on the back of a skunk, and persons wearing such headgear have almost been knocked down by the plummeting attack of an owl.

In these instances, the attacks were clearly cases of mistaken identity; but the great horned owl, given reason, does not hesitate to attack man. Naturalists investigating nests have been severely slashed by the needle-like talons and driven away. Innocent individuals accidentally trespassing upon nest territory have been suddenly beset by a whirl of feathered madness. The great horned owl does not always display this extravagant degree of parental care, but it may at any time. Curiosity with regard to its eggs or young had better be indulged carefully.

An attack is sudden and without the preliminary whirr of wings which might be expected. Instead of stiff pinions, the owl wing is tipped with soft feathers which effectually deaden sound. The normal purpose of this adaptation is to enable the bird to swoop down noiselessly upon unsuspecting small prey.

The great horned owl is a splendid provider for its family and the ordinary nest is surrounded by a stinking profusion of dead game. This, coupled with the ever-present scent of skunk, envelops the nest in an atmosphere easier to recognize than to tolerate.

THE BARRED OWL

The barred owl, *Strix varia*, is now granted the protection of the law. Its food is principally composed of injurious rodents, but on occasion, it has been known to make raids on tree-roosting chickens and on grouse and other game birds, but its value in vermin control should balance these lapses.

The head of the barred owl is round and without ear tufts. In size it is much larger than the screech owl but smaller than the great horned. The coloring is a tawny-gray, marked with the slate-brown bars which give the bird its common name.

The nest is ordinarily the abandoned nest of a squirrel or hawk or the cavity of a hollow tree. Less timid than its relations, the barred owl has no great aversion to man and often nests in woodlands close to settled regions and farms. Even within late years, nests have been found on Staten Island within the official borders of New York City.

The barred owl is gentle and hardy in captivity and, with its own species, amicable. Several may be put in the same cage without fear of feathers flying.

In the wild, the food of both hawks and owls is largely composed of small mammals and reptiles. Birds are indulged in at times, but do not form an appreciably large part of the diet of most species. In feeding captives, whole carcasses of mice, rats and other small game form the best diet. These contain a balance of food elements not present in any one part of the animal body. Even the hair is necessary, especially to the owls, for the proper forming of the cast-up pellets. If meat only is available, it should be rolled in tow or chopped feathers to provide this material. A varied diet is always advisable. Liver, chicken heads, beef heart or any fresh meat, free from fat, may be used to supplement the supply of whole food. When there is a shortage of rats and mice, every fourth or fifth feeding should be dipped in cod liver oil. Proper feedings are a little scant of the bird's idea of what he ought to have, and mealtime should find him keen and hungry. One feeding a day is enough and even this should be omitted once a week. Clean, fresh water in a container large enough for the bird to bathe in if the notion strikes him, should be provided.

Cages for hawks and owls need not be large but must be well-ventilated, well-drained and clean. If cleanliness is not attended to, the premises can be expected to smell rather strong. Flying room is not necessary. A cage with all four sides open to view and wind is bad. The need for keeping watch in all directions at once is too much of a nerve strain on birds and definitely upsets them. This factor, as well as the avoidance of drafts, makes advisable cages with screen fronts only, the other sides and the roof being wind and rain tight. Perches are necessary of course.

In handling owls and hawks, great care should be taken to avoid scratches and slashes. The talons are not only sharp and able to inflict severe wounds but, loaded as they are with decaying animal matter, are apt to be carriers of infectious organisms. Disinfect all wounds at once with iodine or merthiolate and watch carefully for any signs of inflammation.

The trapping of owls and hawks is the phase of the game most likely to result in injury to the human being concerned, but the wearing of gloves and the skillful manipulation of a six-foot square of heavy canvas will do much to lessen danger.

Although young birds are more desirable, these are not always available. Adult hawks and owls may be caught in ordinary steel traps without damage or pain to them. A piece of heavy rubber tubing is split lengthwise, put on each jaw of the trap and fastened securely in place with electrician's tape. When closed, rubber contacts rubber. There is no heavy and painful bite of steel on the bird's fragile leg and, except for outraged feelings and anger, he is none the worse for the experience. Such a trap is set, but not fastened, on a high pole in the vicinity of a farmyard or on some previously noted roosting place. An anchoring wire long enough to allow the captive to flop down to the ground is essential. Pole traps are so much misused that there is a strong prejudice against them. However, if properly made and set, they need not be cruel.

THE BLUE JAY

Even more than the hawks and owls, the impudent blue jay, *Cyanocitta cristata,* has the reputation of being a thoroughgoing rascal, mischievous, malicious and utterly without morality or principle. Without a doubt, the jay has bad habits. He has good traits also, and it is to be questioned whether he altogether deserves the bad name he bears. His great failing is a cannibalistic liking for the young and eggs of other birds. Against this charge is to be credited the fact that the jays are steady, hard-working destroyers of pest insects. The greater part of their food consists of vegetable material—seeds, nuts, acorns and wild fruit. Examinations of stomach contents demonstrate that, after all, the much-talked-about consumption of birds and bird eggs forms but a relatively small percentage of the whole food intake.

The jay is most attractive and its delicate coloring ranks it among our more beautiful birds. Not content to be bright and conspicuous in color, the jay must make his presence known in other ways. Practically a public character, he is raucous and seemingly always engaged in a row with someone. However, it is likely that the jay is no more destructive and vicious than many other birds. The fact that his devilry is flaunted in the public eye is the thing that really condemns him.

In the fall, a chorus of screeching and frantic squalling often shatters the quiet of the woods. The cause of the tumult may be nothing. Or it may be an owl—a screech owl perhaps—dozing out the heavy day in a sparse thicket. Jays have no liking for owls and, meeting them casually, are not content to pass on silently. The subtle insult of the snub is unknown to the jay. He must speak his mind. His sisters and his uncles and his cousins and his aunts likewise must gather and speak their minds. At the top of their voices, each trying to be loudest, the clan individually and collectively tell the owl what they think of owls in general and of his own despicable self in particular.

Just as a crowd gathers to listen to a row anywhere in the world, so are other birds attracted to the fuss. The branches behind the jays fill with the excitement of a twittering and shrieking gallery.

A good deal of the vitriolic tirade is due to the owl's indifference and lack of response. He may blink a yellow-rimmed eye now and again but refuses to become alarmed or vocal. If the annoyance lasts too long or becomes tiresome, he takes wing and, with a turbulent trail of screaming blue behind him, seeks out quieter quarters to finish his rest. The jays, although daring the worst their voices can manage, are careful not to let their insults draw them within the reach of the owl's needle talons. Too many of their brethren's bright feathers rim the nest of the night raider!

The blue jay is a braggart and normally cautious. However, fear for the safety of the nest and its precious contents sometimes overthrows caution and results in attack even on persons who venture too close.

Ordinarily, jays are too much interested in making unflattering comments upon the behavior of stray humans to give a display of their most engaging trick—mimicry. Unaware of the presence in the woods of an observer, jays have been known to sit on a limb and, in rapid succession, imitate perfectly the calls of a whole string of singing birds.

The jay is a good provider for his growing family during the nesting season, and, during that time, drops all turbulent

U.S. Fish and Wildlife Service
photo by Rex Gary Schmidt

R. Elwood Logan

SPARROW HAWK

BARRED OWL

U.S. Fish and Wildlife Service, photo by Gale Monson

YOUNG GREAT HORNED OWLS

U.S. Fish and Wildlife Service, photo by H. H. T. Jackson

CROW

U.S. Fish and Wildlife Service, photo by E. R. Kalmbach

MAGPIE PILFERING EGGS

habits and devotes the day to hunting food and standing guard.

In captivity the blue jay is noisy, intelligent and amusing. In a household, unless stowed for the night in a dark place or well covered, with the dawn comes awakening—for everybody.

THE CROW

Just as the hooting of the owl is considered a mournful and forbidding note, the repeated "caw" of the crow seemingly has in its harsh and raucous tone an undercurrent of derision and ridicule. Actually it has not, but it well might have. Detested and disliked, these midnight black and noisy thieves add insult to every injury and by a perverse intelligence foil most attempts to control their activity. The impotent farmer, unable to get within shotgun range of the flock ripping out his sprouting corn, can only shake his fist and set up in his field that ancient and comical device, the scarecrow.

The crow is a wise bird and, very literally, one has to get up early in the morning to fool him. Depredations in the neighborhood of farms or houses are usually committed during the hours when lie-abed humans are still asleep. In addition to raids on the fields and gardens, crows attack small chickens and the like. Eggs of fowl and bird are one of their fancies, but it may be noted that they do not spare the nests of other predatory birds such as the hawk and the owl.

Intelligent and canny, the crows make a very successful job of living and, in spite of a long history of persecution and dislike, their numbers are probably greater now than they ever have been. Thoreau speaks of them: "This bird sees the white man come and the Indian withdraw, but it withdraws not. Its untamed voice is heard above the tinkle of the forge. It sees a race pass away, but it passes not."

The nests of crows are usually—but not always—in woodlands. Well-constructed, sturdy, somewhat sloppy-looking from below, they are well lined with all sorts of warm and soft material: rags, grasses, seaweed, hair, soft roots and the like. The height of the nest from the ground depends upon the trees available but is sometimes in excess of sixty feet. As many as nine eggs have been found in a nest, but in these cases it was assumed that two females shared the same nursery. The usual brood is from three to five.

Young crows make splendid pets. The household which

belongs to one never lacks for excitement. When it is a baby, there is the unremitting demand for food; when older, if at liberty about the premises, a succession of wild escapades and mad thievery keeps the house alert. Things disappear. Spoons, jewelry of all sorts, small kitchen tools, pipes—any small object the crow can carry—may be taken. He has no particular use for them. He just wants them. The articles need not be irrevocably lost. The thief is himself suspicious and checks up on his hiding places whenever he has nothing else to do. Frequently he removes the cached treasure to a second site and, if his movements are watched, can be caught with the goods. As might be expected, he sets up a loud protest when despoiled of his loot.

Tame crows often become very much attached to individuals and follow the chosen master or mistress about, making all sorts of bids for attention. Squeals, laughs, pecks on the ankle, whistles and plain malicious mischief are some of the means used. The crow may insist on being allowed to help with household tasks or in the garden. If his efforts are not appreciated and he is sent about his business—watch out. His business at the moment will be to get even for the slight and he'll attend to it!

While not notable linguists, crows frequently manage to pick up a few words and learn to imitate the commoner noises of the household. The old idea, that if the tongue of jay, crow or magpie is split the bird will learn to talk, is still current. This is rank nonsense. If anything, such an operation would be much more likely to make speech impossible.

The food of crows in nature is varied. Domesticated and allowed at liberty, the diet may be almost anything that comes to hand. In a cage, with less exercise, it must be a little more strictly supervised.

THE MAGPIE

The magpies of the West are closely related to the crows and jays and in many respects very much like them. Extremely adaptable to circumstance and able to get the best out of it, the harsh-voiced magpie, like his relatives, has earned much dislike by the open and derisive manner in which he outrages man's dignity and property.

The lustrous black of his upper plumage is iridescent in the sunlight and admirably contrasts the white of the underparts. Larger than the crow, the magpie is a much handsomer bird. The species found east of the Sierra Nevadas and throughout the Rocky Mountain region, *Pica pica*, has a black bill while the California species has a bill of bright yellow.

In captivity, the magpie has all of the amusing and distressing habits of the crow. He is a braggart, a thief and a liar; a black and white bundle of antics and alibis. Decidedly not a shy or diffident bird, he insists on everyone knowing what he is doing and on knowing what everyone else is doing. Much more fluent than the crow or jay, magpies learn to talk fairly well and in addition to single words, chuckles, whistles and maniacal laughter, add whole sentences to their repertoire.

The jays, magpies and crows are not birds to be confined in small quarters. They need at least the run of a small flying cage or, better still, if thoroughly tame, may be left at liberty with one wing clipped to keep them from straying far. Of course, this is not possible is nesting birds are to be encouraged or if there are open enclosures for reptiles, frogs or small animals in the yard.

The cage should have all sun possible and be well drained. These birds are hardy and able to withstand severe cold if well fed, but quickly succumb to damp. Dry shelter in which they are protected from driving wind and rain is imperative. Perches and branches for roosting are necessary. This group of birds is emphatically not for the community flying cage. Murderers all, no one of them can be trusted with anything he can kill.

They are not particular about food. Coarse, soft foods with meat scrap cooked into them are good. The canned dog food now on the market with its mixture of cooked grain and meat is excellent. Avoid feeding hard, dry grain. If used, cook or soak it. A bone with a few tag ends of meat hanging to it furnishes not only nourishment but amusement. For the crow and magpie, mice form a valuable addition to the diet. The food of jays indoors in small quarters needs a little care. Prepared mocking bird food fed with a supplement of nuts, seeds and occasional insects is satisfactory. Clean water for drinking and bathing must be available at all times.

The hand-rearing of young birds is a chore which should be well considered before it is undertaken. Preferably, youngsters should not be removed from the nest until they are able to feed themselves. Unfortunately, things do not always work out that way, and the impetuous collector may find himself with a baby case on his hands. He must remember that the most important thing in the life of a young bird is food. His mind is always on it and visitors are immediately greeted by a gaping and hungry mouth. The rate of growth of young birds is so great that meals have to be practically continuous. Young crows normally eat half their weight each day they are in the nest and have been known to consume even more. This bulk of food cannot be given to them at one feeding but must be staggered throughout the day. If a young crow or other bird is to be brought to a healthful, full-sized maturity, his owner may as well make up his mind to stay close to the infant and fill that ravenous maw several times each hour. Large feedings hours apart will not do. Small feedings at frequent intervals approximate the natural feeding manner and periods. As the bird grows and becomes able to take care of himself, these irksome duties slacken and eventually become unnecessary.

The food of these youngsters should be soft, almost sloppy, and lukewarm. Recipes vary with the preferences of the feeder and the materials available. Ordinary bread and milk or zwieback soaked in milk are frequently used. A better-balanced ration can be made by mixing dog food with rice, ground carrots or hard-boiled egg. A little cod liver oil should be added. Another good food is made of a beaten raw egg, two teaspoonfuls of ground puppy biscuit, a pinch of calcium lactate, a couple of drops of cod liver oil and enough raw chopped beef to stiffen the mass. This mixture should not stand out of the ice box too long. Naturally, it must be warmed slightly before feeding.

Well-intentioned persons with no idea in the world of bringing up foundlings, sometimes afflict themselves unnecessarily with the care of a young bird they have picked up. The waif is always assumed to be homeless and helpless. Actually, these babies are seldom the deserted orphans they seem to be. The mother is usually somewhere in the offing and, if the young are let alone, will come and take care of them. Indeed, there are recorded cases in which females of other species, finding such young, have hunted food for them and watched over their welfare. Unless the individual is prepared to be responsible for a tedious and often futile job, he had better pass quietly on. The best service he can render and the way in which he can really help to insure the survival of the young-sters is to see that no cats, jays or other prey-seeking creatures venture too close to the vicinity.

As with other creatures, keeping up birds' resistance to dis-ease by feeding and care is easier than the treatment of dis-ease. It is difficult to do much about serious sickness and it is usually up to the bird to save himself. The owner can best aid by providing even and constant heat, fresh air and a freedom from drafts.

Parasites infesting cages and birds are a constant problem. The cages can be taken care of by a good, thorough cleaning. The birds themselves are a more difficult matter. Many mite remedies have possible toxic effects and cannot be used safely. Birds heavily infested with parasites lose interest in food and become listless and thin. Danger is also present from parasite-carried disease. A recently marketed preparation, called Dri-Die, is a silica serogel dust insecticide which kills insects by dissolving the thin wax coating that retains the body fluids. Treated insects lose great quantities of moisture and die with-in minutes. The very light powder is best applied to the bird with a duster. It may also be used to delouse corners and cracks of the cage.

A number of species of game birds are now being bred for stocking game preserves and estates. The young and eggs of these can be purchased from recognized dealers. These men have full information on the housing, feeding and breeding of the species they handle and, in addition, are able to say ex-actly what permits are necessary in a particular state. The advertising pages of such magazines as *Field and Stream* offer many suggestions for interested persons.

BIRDS, DDT AND THE REST OF US

Chemical warfare is a two-edged weapon and its use in the conflicts of humankind were banned after World War I as be-ing too dangerous to all parties concerned. However, in man's fight with the world of undesirable insects and plants, its use is becoming alarmingly widespread. The weapon is still two-edged and the present wielder of it is now beginning to find himself suffering severe damage from his own instrument of destruction.

Through the ill-advised and careless use of sprays to combat such pests as flies, mosquitoes, and carriers of some plant diseases, many areas formerly rich in small wildlife—insects, reptiles, amphibians, as well as birds—have become biological deserts.

The sprays have been far from satisfactory for the destruction they were intended to work but have been all too effective in wiping out these small innocent bystanders. The insect pests themselves are notoriously tenacious of life and able to make adjustments to adversity that are fearful and wonderful. The fly and mosquito have been able to survive our chemical warfare and develop strains which are not only immune to our poisons but seem to thrive on the dosage. The nuisance persists and, in the meantime, desirable kinds of life have been badly damaged.

In sprayed areas, birds—robins, bluejays, cardinals, thrashers, woodpeckers, and many others—are picked up daily, dead or dying, with the typical symptoms of spray poisoning.

Birds are not the only victims. Such ground-dwelling insect eaters as shrews, frogs, toads, and salamanders have almost entirely disappeared from some areas. In the ponds and streams, fish and the horde of small aquatic creatures that normally supply fish with food have been exterminated. The moths and butterflies, so long one of the great joys of our fields and woods, have been terribly hit in sprayed areas. Today, in some places, all a publicity hound has to do to get his picture in the papers is to find a Luna or Cecropia moth, once so common, to pose with him.

Effects can be far-reaching. In an experiment of the early days of spraying, a roach was poisoned with DDT and fed to a hungry mouse. The mouse, starting to show signs of distress, was given to a corn snake. Several days later, after the snake's slow digestion had rendered the poison available to its system, the typical symptoms appeared and the snake died. The weed-killing preparations, like the insecticides, may have wide ranging and unexpected side effects that should, perhaps, make us give a little selfish thought to our own welfare.

Proponents and practicers of these programs, even after their efforts have been demonstrated to be not only futile but risky, have been known to justify themselves by saying, "Well, we figured we had to do something and this seemed like a good idea at the time."

If, as has been sourly and sadly suggested, man, by the use of his great cleverness and inventive ability, manages to make this earth completely uninhabitable for his kind, perhaps a fitting inscription for a monument with no one left to erect it might be:

IT
SEEMED LIKE
A GOOD IDEA
AT THE TIME

The Terrarium

EMBRACING AS it does all sorts of containers from pickle jars to architect-planned, elaborately landscaped setups, the term *terrarium* may at first sight seem hardly worth the using. But it is the only word we have to indicate that through the use of the pickle jar, the architect's expensive creation or something between the two, an attempt has been made to establish a community of plants or of plants and animals in a comfortable and natural association.

The pickle jar, at any rate, has a clear title to the name. The story goes that Dr. Nathaniel Ward, the nineteenth-century English physician who invented the terrarium, one day put a moth or butterfly cocoon in a glass bottle to hatch. A week or so later, the good doctor noted inside the bottle, indifferent to the prevailing heat and drought, a lusty growth of young ferns and other plants. His pondering over the secret of the bottle led to the development of the Wardian case, with its display of exotic plants—a most highly prized and admired feature of the fashionable Victorian parlor. It is the secret of the doctor's bottle—controlled heat, light and humidity—that makes the terrarium, our modern version of the Wardian case, an ideal home for the reptile and amphibian guests of the small zoo. The arrangement and planning of these little worlds is a work particularly suited to the interested individual or the nature room.

Zoological parks more and more make a practice of exhibiting their larger animals in natural habitat settings, but seem to find it convenient to show the smaller creatures in bare cages or tanks. While affording a good clinical view of the animal's anatomy, if the visitor is lucky enough to catch him at a moment when he is not huddled into a compact, scared-looking lump, such exhibition fails to give as much information about the animal's character and living habits as a well-mounted museum group or a photograph does.

This is the small zoo keeper's opportunity. It sounds like a large order—showing a healthy and growing animal in his natural setting together with his accustomed plant and animal

TERRARIUM

associates—but with the reptiles and amphibians, it can be accomplished within the narrow limits of the terrarium.

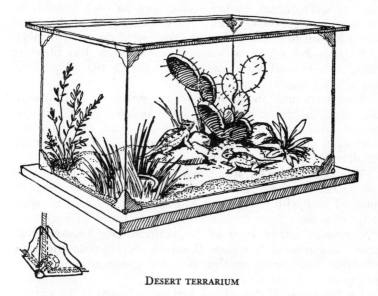

DESERT TERRARIUM

The most satisfactory type of terrarium for animals is something along the lines of a fish tank—a rectangular container with four glass sides, a watertight base if possible, and a glass or screen top. A discarded aquarium is splendid. If such a tank is not available, four panes of window glass may be fastened together with adhesive tape and set in a shallow baking pan or tray, and another pane of glass used for a cover.

An extremely practical method of making a terrarium has been devised by Dr. Forman T. McLean, formerly of the Bronx Botanical Garden. Although originally intended as a plant container, this setup is particularly well suited to the zoo because of its attractive and modern appearance, its flexibility and its low cost. It consists merely of four panes of window glass clamped together at their corners and set upon a wooden base. The clamps are made of two three-sided trunk corners of the type easily obtained in large hardware stores. A pair of these, drawn together by a brass screw through holes bored through their centers and held by a hexagonal nut, make an efficient corner to hold the panes of glass at right angles to one another. Thin cork or other material cemented to the inside of the clamp serves to cushion and protect the glass when the screw is tightened. In some sizes of terraria, a trunk corner outside and a heavy rubber washer inside may be used instead. In order to insure a close fit, the corners of each pane of glass must be nicked off to give room for the passage of the screw through the clamp. For a base, a two- or three-inch deep tray of white pine, well painted to guard against warping and, if possible, metal-lined, is quite practical.

This terrarium, in addition to its simplicity and modern lightness, has the advantage of being knock-down and can be stored easily when not in use. By this method, too, large size enclosures may be made which would otherwise be outside the small zoo or nature room budget.

The success of a terrarium is in no way dependent upon its size or elaborateness. A small battery jar or fish tank, properly planted and with a carefully edited guest list, may be more attractive and *right* than an expensive, badly planned and hodge-podge affair. It must be remembered, however, that small and precious creatures thoughtlessly put into the same container with larger relatives often, as guests at supper, turn out to be the honored dish.

For the health of the animals and the rightness of the picture, every effort should be made to establish natural plantings in thorough keeping with the resident's needs and character. Pigs do not belong in parlors and there is something equally ridiculous, and, at the same time, pathetic in the sight of an Arizona horned toad crouching under a fern on a damp wad of sphagnum moss. The effect of desperation is heightened if a cactus is slowly dying of wet feet three inches away.

With its steady food supply, its control of moisture and its freedom from the menace of the heron, the water snake and the baker's dozen more of persistent enemies, the terrarium provides a safer and more comfortable home for toads, frogs, salamanders, many lizards and small snakes than is vouchsafed the average animal in the wild. For such small creatures these narrow bounds are not too rigorous a confinement. Indeed, within these limits, the whole cycle of reptile and amphibian life may run its course.

Roughly considered, the small creatures adapted to terrarium life may be divided into several large classes: woodland, marsh-stream and semi-arid or desert. Naturally, to receive a member of any one of these groups the terrarium must be a bit of that particular setting brought indoors.

THE WOODLAND TERRARIUM

With the woodland terrarium, this is best done in very literal fashion. Start out with a strong cardboard box, a trowel, a heavy knife, a newspaper and a couple of paper bags. Head for the wildest stretch of woods you know. Look about and catalogue the possibilities. Moss? Take it up very carefully in large sheets or in the rounded hummocks some species form and put a layer face down in the bottom of your container. Try especially to get clumps of moss that already have small plants rooted in them—partridge berry, Canada mayflower or the like. Put the next layer of moss in face up thus placing the two dirt surfaces together. Cover over with a layer of paper before laying in the larger plant material. This should be carefully dug out—not simply uprooted—together with the mass of earthmold enveloping the root. Woodland plants such as the gentians and the trailing arbutus are not only too delicate to withstand the shock of even a very careful transplanting but are protected by law and may not be taken. There is, fortunately, an infinite variety of small plants, equally attractive and better suited in scale to the miniature landscape, which, with reasonable care, transplant readily.

Wrap the selected plants *loosely* in paper, being especially careful not to shake the dirt from the roots, and put into the box on top of the mosses. Look about the woods before you go, for a few decorative accessories—bits of fungus-laden wood, half-rotted and gnarled pieces of root, lichen-covered bark and rock. Don't overlook a few dried leaves.

These are the things that give the touch of reality to the terrarium landscape. The paper bags are the convenient containers for a generous helping of the rich, black leaf mold bedding the woodland floor.

Planting the terrarium comes next. There are people with what the Irish gardener calls the "green touch" who seem to be able to take a little of this, a little of that—assorted odds and ends of moss and plant material—and in no time at all toss together a good-looking and practical arrangement. We are not all so fortunate and setting up the terrarium must be, for most of us, a matter not of inspiration, but of calculation and careful judgment.

The first step is the preparation of the container. In order to prevent any sourness of soil, an inch of charcoal or cinders, held down by a layer of pebbles, or, better yet, a layer of broken flower-pot shards, makes a good drainage layer. The front of the container is the frame of the picture-to-be and, as a cross section of charcoal, broken pots and earth is hardly decorative, the drainage layers should be sloped from nearly nothing in front to several inches deep in back. Or, if your terrarium is to be very large and its contents visible from all four sides, consider all these points of view and slope from sides to center.

Cover the drainage layer with a couple of inches of well-sifted topsoil of a somewhat sandy texture and over this spread the leaf mold.

As in any landscape, real or painted, the effectiveness of the terrarium picture depends upon the disposition of the larger masses. If the accessories and planting are all flat on the bottom of the tank, the materials merge into a monotonous, single plane effect. Differences of level, important in any landscaping, are especially valuble in the terrarium for the feeling of added depth they give. In the building up of these differences of level, rockwork is one of the most useful and certainly the most abused of terrarium features. In most cases, one large, irregularly shaped rock or a compact group of smaller stones embedded in the soil high at the back of the terrarium is sufficient not only to hold the higher parts of the planting in place, but also to give the effect of a natural outcrop or ledge.

In selecting stones, discard those too regular in shape and those very strong and bizarre in color. Avoid using two widely different types of rock in the same setup. In particular, relegate to the what-not in the parlor that clump of rose quartz crystals and the pieces of gem-like petrified wood. Or—if you feel they must go into the terrarium—bury them so deeply under the earth and plant materials that not one glint from their polished surfaces betrays their presence. Accessories of this type strike too positive and, in this case, discordant a note for the quiet harmony of the woodland scene and can never appear a natural part of it. In placing rockwork and any large pieces of stump or wood, see that little crannies and pockets for planting are left.

At this point it might be well to spread out *all* the plant material in full view, even at the risk of its drying out a little. If this is not done, some particularly attractive bits are sure to be overlooked until after the planting is finished.

The larger rooted plants should be taken care of first. The seedling evergreens—pine, hemlock, spruce or cedar—will find root room in the deeper soil at the back. One tree will be enough for most setups. Next, use the pockets formed by the rockwork to hold ferns and smaller plants. It will very likely be necessary to shift material from place to place several times before just the right effect is produced. When the arrangement seems pleasing and natural, tamp the earth about all roots. If the plants have been set in much the same relationship in which they were originally found, the sheets of moss, firmly pressed into place, each with its spangling of tiny seedlings, will fit into the space between with no sense of overcrowding.

If you have been fortunate enough to get a few trailers of the partridge berry with its scarlet fruits, or of teaberry or other small vine which will root from cuttings, pieces of it may be planted in holes jabbed through the moss with a pencil.

A few dried leaves, a fungus-laden stick or so, perhaps a piece of bark, dropped in carelessly but artfully, complete the picture. All that remains is to spray well—but not too well—and the terrarium is ready for its woodland guests.

THE MARSH-STREAM TERRARIUM

Swamps, marshes and the borders of streams are the natural haunts of the frogs, turtles and salamanders which are so prominent in the list of possible guests of the small zoo, and it is usually necessary to reproduce this setting in a varied series of arrangements to accommodate them.

To be most satisfactory, this type of terrarium should be made from an aquarium for the sake of a clear view of the underwater life. Both land and water sections are needed and, for the good of the plants on the artificial stream bank, water must be kept out of the land area. A practical method of doing this is to divide the rectangular tank lengthwise or diagonally by a strip of plate glass less than half the height of the tank. The partition should be well fastened in place with aquarium cement or with plaster of Paris tinted earthbrown with dry burnt sienna or umber. If plaster is used, let dry thoroughly and shellac well. When the partition is reasonably waterproof, put an inch or more of powdered charcoal and broken pottery on the side to receive the planting of dry ground material to ward off any tendency to sourness. Fill the space almost full of good, rich sandy loam and the bank is ready for its planting of small ferns, sedges, marsh-marigold and other water-side habitués. Rocks, moss-covered logs and other natural accessories may be piled at the back, as in the woodland terrarium, to add perspective and character to the scene. Pebbles, piled-up sand and aquatic plants should be arranged in the water section to conceal the partition and to provide a natural transition from the water area to the land, as well as to give to creatures living a double life, a means of passage to and from the higher level.

It is obvious that the variations on the marsh-stream motif are, as in that of the woodland, limited only by the number of containers available and the builder's ingenuity, patience and observation. In every two feet of woodland or of pond or brookside, there is a beautiful terrarium waiting to be moved inside.

As far as the plant materials of both the woodland and the marsh-stream terrarium are concerned, partial sunlight and a cool location are best. Fortunately, most of the animals suited for residence in them prefer these conditions. However, there are exceptions in the marsh-stream group, and for them the terrarium needs full sun for the greater part of the day. In such cases, a wire screen cover must be substituted for the tight glass top, to allow circulation of air and to keep the terrarium from overheating. If mildews and molds appear in the woodland terrarium or if newly sprouted ferns and young plants "damp off" and die, the condition can be remedied by substituting a screen top for the glass cover for a day or so. After surplus moisture has been allowed to evaporate, replace the glass and watch watering more carefully. If the drainage layer is adequate, this trouble is not likely to occur.

Moss for replenishing the terrarium during the winter may

be gathered in late fall, wrapped in wax paper and packed in cardboard cartons for storage in a cold place until needed. Plants may also be kept for a time in this fashion but are, of course, less long-lived than the mosses. Last year, I was blessed in early spring with a collection of pine barren plants: pitcher plant, sundews, pixie moss and bearberry; instead of planting them immediately and allowing them to come into bloom, I put them in the ice box (50°F.) and left them. Two months later they were taken out and planted for a special exhibit. In response to a couple of days' warmth and sun, the pixie moss and the bearberry put out a delicate spangling of bloom. The sundews unfurled their slender red-haired leaves and shot up a flowering stalk. The pitcher plants added new green cups to their clusters of flycatchers and the nodding, sculptured flowers, held high above the damp mosses, assumed the rich red of maturity.

THE DESERT TERRARIUM

It is possible also in addition to holding back bloom, to grow from seed and force into flowering at a particular time a great many attractive wood and field plants.

Seeds of many of the western arid and semi-arid plants germinate quickly under the influence of a little heat and water and make a speedy growth and display. Many of these may be sown in place in the terrarium or, better yet, should be sown in shallow pots and the pots, after the plants are well established, embedded in the soil. This latter practice permits the easy replacement of torn or out-of-bloom plants from a reserve stock. Although the arid terrarium is intended to replace the conditions and climate of dry and semi-desert regions, the first step in its construction is, as in the case of the woodland terrarium, the placing of an adequate drainage layer. The cacti are the plants most characteristic of deserts and although built to store up moisture, soggy soil and wet feet mean inevitable destruction to them.

For the benefit of these plants, which thrive best in an alkaline soil, mix in a little slaked lime with the sandy loam that is to cover the drainage layer of broken pottery or cinders. A better plan is to keep the cacti in small pots of lightly limed soil and embed these in the soil of the terrarium bottom. An air of reality can be achieved by the addition of grass—not a smooth and darkly green stretch of sod, but tumbled and wispy grass,

starved for moisture. Like the other plants for the arid terrarium, the grass is best handled planted in its own shallow pot
or in a flower pot saucer. In this fashion, each plant may be
given its own appropriate portion of water and the humidity of
the whole not affected. The clumps of grass are of special value
in masking the corners of the terrarium and to hide the edge
of the water container. Cacti suffer from sprinkling but the
grass welcomes a daily spray from the atomizer and the drops
sliding down the stems and drooping blades offer a welcome
watering place to lizards unaccustomed to drinking from a dish.

Although in making a terrarium of any type every attempt
should be made not to mix plants of one region with animals
of another, there is no need to be precise about if it the picture
suffers. After all, the effect of a natural setting is the end desired. If a clump of bedraggled grass from a New England fence
corner must suffice to take the place of the bunch grass of the
prairie, use it. The only fault in such a substitution would be
in the use of an ornamental grass with striped and variegated
foliage or of grass too green and lush in character.

Rocks and accessories for the arid terrarium should be selected with care. Obviously, a moss and lichen-encrusted bit of
ledge is entirely out of key with the animals and plants. Desert
animals have developed protective coloration to a high degree
and the smaller reptiles, especially, tend to assume the general
hue of the environment. By matching the basic coloring of
rocks and sand to the general tone of the animals and keeping
them in accord, a right effect will certainly be achieved.

Nature is untidy and in the midst of life, leaves scattered
fragments and relics of death. If you have a bit of dead cactus,
rotted and hollow, copy nature and put it in. It is not impossible that the composition and space might make logical and
proper the inclusion of even a sun-dried and bleached bit of
bone, a broken and splintered hub of an old wagon wheel, or
any other small and pertinent bit of debris.

The Amphibians

THE LIFE histories of frogs, toads and salamanders, although varying in detail, follow the same general pattern. The term *amphibian* under which they are grouped applies not only to the semi-aquatic habits of most of them, but describes as well the curious larval or tadpole stage of their life history which sets them apart from the other cold-blooded backboned animals. With but few exceptions—chiefly salamanders—the life of our amphibians begins in the water, although in maturity many of them leave it to take up a terrestrial existence.

With our native frogs and toads, mating and egg-laying take place in the water. The egg-heavy female is attracted to the vicinity of the male by his calling. Then, clamping his forelegs in a tight grip about her middle, he hangs to her back—sometimes for many hours—until she has voided her clutch of eggs and he has poured out the fertilizing sperm upon them. The egg-laying accomplished, the parents promptly go off and forget about their young. After a few days of incubation (the time is somewhat dependent on temperature as well as species) the tadpoles, or pollywogs, as they are sometimes called, emerge from the large gelatinous egg masses. Longtailed, fish-like, scavenger-vegetarian offspring, they bear hardly any resemblance to their tail-less, carnivorous, air-breathing fathers and mothers. Fortunately, this aquatic condition does not persist. The time taken for transformation from the tadpole stage to the adult form varies greatly with the several species, but, sooner or later, the tad develops hind legs; the swimming tail begins to show signs of wear and tear and shrinks; front legs pop out and the thin, transparent skin through which the long watch-spring-like intestine could be so plainly seen takes on pattern and color. Finally, some fine morning, a tiny frog or toad, as the case may be, with just the hint of a tail, sits on a leaf or rock waiting for something small in the way of insect life to furnish him with his first flying breakfast.

If ponds are visited in the spring, especially at night, the males' chorus of calls makes frog and toad collection easy, and a clasping pair that has not already laid is seldom difficult to capture. Normally shy and evasive, these creatures are almost entirely oblivious of danger while mating and make none of their customary frantic efforts to escape. Such a pair, placed in an aquarium, readily take advantage of the chance to lay their eggs. For better development and a clearer view of the hatching of the tadpoles and their metamorphosis to adults, it is best to remove the eggs to well-planted battery jars or tanks. The parents should either be turned loose near the spot where they were captured or given a home in a suitable terrarium.

Tadpoles, being largely vegetarian, are easily taken care of. Lettuce, well cooked, should be dropped in the tank in *small* quantities. An occasional bit of raw liver proves an appreciated delicacy, and quickly disappears under the onslaughts of the youngsters.

After metamorphosis, young frogs and toads are apt to drown in the deep water of the aquarium unless there are floating plants or debris on which they can climb out. At this time, they are ready to go into a terrarium, but, in the case of frogs especially, it had better not be the one in which their parents have their residence. There is no greater connoisseur of tender young frog's legs than a bigger frog.

A characteristic common to the amphibians, and one which serves to distinguish them from the fish and the reptiles, is their scaleless skin and the part it plays in their lives as an organ of respiration. This is especially true of frogs and salamanders. During the cold winter months when they lie buried in the muddy bottoms of ponds or under the stones of streams, the skin, rich in small blood vessels, offers the only avenue by which the slowly circulating blood may exchange its burden of carbon dioxide for life-maintaining oxygen. In warmer weather, when the frog has come out of hibernation and is at the surface or on land, the animal is still greatly dependent upon the skin as a breathing medium, although its lungs function. If skin respiration is checked through lack of moisture, the frog perishes. It is this dependence upon moisture which keeps the smooth-skinned frogs, the tree toads and the salamanders confined to the immediate neighborhood of water or in moist woods.

THE TOADS

The toads have partly freed themselves from bondage to water through a more efficient internal respiration than that of the frogs and by the development of a rough and warty skin which greedily takes up water but is chary about letting it out. This rough warted skin in still another fashion gives the toad a freedom to travel not granted to the other amphibians. The warts, particularly the two large bumps immediately behind the eyes, are actually groups of glands which exude a gummy, white fluid upon the seizure of the toad by a larger animal. These defensive skin glands, although present in all the amphibians in some degree, reach their highest development in the toads. The fluid is irritating in the extreme to the sensitive mouth parts of other creatures. However, it by no means renders the toad invulnerable or undesirable. Small toads are the victims of predatory birds and even the domestic duck and hen. Some snakes, particularly the hognose and garter snakes, find a dinner of toad acceptable. Curiously enough, unlike the frogs, the toad is seldom cannibalistic—though accidents will happen.

Under ordinary circumstances the fluid is harmless to man, but, if through carelessness it gets into the human eye, severe inflammation is likely to follow. Contrary to popular belief, the secretion does *not* cause warts.

City-bred dogs, and young puppies unversed in the practical matters of life, are likely to get into difficulties with the first toad they meet. The memory of the miserable and painful time following this encounter invariably builds up an extraordinary politeness and respect for the species.

This meager defense is all that is left to the toad of that legendary store of venom and ill will with which the folklore of many races credits him. Too, he has lost his wealth. The jewel he wore in his head, the precious toadstone, has been stolen and he is left with the gold and black beauty of his eye for sole ornament. Time has converted him into a quiet country gentleman, an inoffensive person in reduced circumstances, not at all handsome, but courteous and quiet and most neighborly to man. This friendship to man is becoming more and more a matter of official record. A very active citizen, the toad has been estimated by some authorities to devour more than three hundred harmful insects a night during the warm summer months.

Hidden during the day under buildings, in woodpiles and other moist, cool places, the toad emerges at dusk to go about the night's business. Although hardly built for speed, he can display an unexpected liveliness in his nocturnal hunting. His ordinary method of locomotion is a series of short hops but, stalking a caterpillar, worm or beetle, he often resorts to a fast crawl, a gait less likely to alarm the prey. Most animals find it necessary to approach their victims closely enough to grasp them with claws or teeth. Not so the toad. For that matter, he has neither claws nor teeth. He has a tongue which takes the place of speed, claws and teeth. Long and covered with a sticky exudate, it is attached at the very front instead of the back of the mouth, and in a lightning-quick movement may be flicked out to capture an unwary insect.

We have many species of toads within the United States, all of them more or less similar in general appearance and usefulness. There is considerable difference in size, degree of wartiness and voice, however. The small oak toad of the south-eastern states, *Bufo quercicus*, a midget, grows to little more than an inch in size, but congregates in numbers in ponds and pools for breeding, and is prodigious for noise. The shrill, high-pitched piping, multiplied by thousands, is ear-splitting at close quarters.

Almost every area has its familiar and friendly species. In the eastern states, *Bufo americanus*, the American toad, prowls through every garden. In the south, *Bufo terrestris*, very much like the American toad in character, appearance and usefulness, appears at dusk to reap the insect harvest. The stocky and clumsy California toad, *Bufo boreas halophilus*, drags its bulk through the rich farmlands of the valley and the high mountains. Even the desert has its toad. The very large Colorado River toad, *Bufo alvarius*, although a dweller in desert areas, is more aquatic than his relatives and makes his home in the vicinity of irrigating ditches and reservoirs and the larger streams of the Southwest. Unlike other toads, *alvarius* has a smooth green leathery skin with but few warts. However, the skin glands are present in full measure and potency, and reliable sources cite instances of dogs having died of poisoning after seizing one of these otherwise inoffensive batrachians.

THE SPADEFOOT TOADS

In addition to the toads of the genus Bufo, we have in the United States the spadefoot toads of the genus Scaphiopus. These are seldom seen, although they may exist in a region in considerable numbers, and are heard but once yearly for a brief season. Burrowers in the ground, they are able by means of hind legs especially adapted for digging to sink from sight in

C. M. Bogert

SPADEFOOT TOAD

Cecil Hartson, American Museum of Natural History

THE TOAD *Bufo terrestris*

U.S. Fish and Wildlife Service, photo by W. P. Taylor

COLORADO RIVER TOAD

GREEN TREE FROG (left)
GRAY TREE FROG (right)

C. M. Bogert

LEOPARD FROG

Robert Snedigar

BULLFROG

C. M. Bogert

PIGFROG

C. M. Bogert

a very short time. Their hunting and feeding time is in the night and like cats, they have good vision in the dark. Instead of having a round pupil in the eye, like most nocturnal creatures the spadefoot has a vertical pupil which may be widely extended to admit all possible light. These toads are apparently very irregular in their emergence from their burrows, and unlike the common toads, do not hunt nightly. Occasionally, in plowing, they are turned out, but, earth-covered and clod-like, usually escape notice.

In the spring, when temporary pools and puddles dot the countryside, the spadefoot toads court and mate, and make their presence in a locality known by their din. If their number is large during the few nights they remain in the ponds, the noise is astounding and effectually disturbs human neighbors. The call is harsh and strident and carries a considerable distance. Although it is to be doubted that they were much noisier than the common toads, there was great complaint from Connecticut towns several years ago that the spadefoot of the East, *Scaphiopus holbrookii*, was making such a racket that no one could sleep. The spadefoot of the West, *Scaphiopus hammondii*, uses spring rain pools and even roadside puddles as its breeding place. Its noise, although loud and perhaps harsh and raucous to persons not caring for amphibian chorales, is less strident than that of its eastern and southern brethren and, softened a little by distance, not unpleasant.

Because of their nocturnal and burrowing habit, the spadefoot toads are not ideal creatures for the terrarium. Nevertheless, their unusualness and the fact that they are so seldom seen, warrant anyone in trying to keep them as an exhibit to be dug out on special occasions.

Although the toads are by no means as free from bondage to water as the lizards and snakes, they are able to withstand degrees of dryness which would cause the thinner-skinned frogs to shrivel up and die. The latter are more brightly colored and graceful in movement than the clumsy and protectively colored toads. Of them, the family of the Ranidae is the largest group in number of species found in North America, and lists among its members almost all our aquatic and ground-dwelling frogs.

THE LEOPARD FROGS

Because of their wide distribution, the leopard frogs, *Rana pipiens* and its allies, are perhaps the best-known of all the tribe. Many students who have not had the privilege of meeting them on their own ground have been introduced to them in class. The leopard frogs are those most commonly used in the zoology laboratory for study and dissection.

In person, the leopard frog is a conspicuously beautiful animal. The olive-green or green body color is heavily blotched with the creamy yellow dark "leopard" spots and the whole overlaid with a shining bronze iridescence. The underparts and vest are a fine-textured ivory white. The color is subject to wide variation but follows this general scheme. The skin is smooth except for the two great folds running from the eye to the pelvic regions. The general impression the frog gives is one of speed and activity. The southern leopard frog, with its more pointed head, slender body and long slim legs, especially gives this impression.

Hibernating during the winter in the mud of ponds or underneath the debris and stones in the bottoms of streams, the leopard is one of the first frogs to emerge in spring and start the chorus heralding mating and egg-laying. The voice of an individual is not loud—merely a low, guttural croaking—but when a large number are calling in a pond, the noise is considerable and its character, pleasant or unpleasant, is entirely dependent on the hearer's feeling about it.

During the several weeks in early spring when they are courting, it is not difficult to observe the making of the leopard frog's music. Sitting in shallow water, half-submerged, the male frog can be seen to take a big breath of air. A swelling appears over each shoulder and the croak begins. As the swellings grow larger, the croak gets louder until the whole frog seems to vibrate with his own noise. Kept distended for some moments, the vocal sacs collapse at the end of the song with all the suddenness of a pin-stuck balloon.

The eggs of the leopard frog are laid in masses of four to five hundred, attached to sticks or reeds along the edges of the ponds or marsh pools. The tadpoles metamorphose in late midsummer and at this time, great numbers of the young frogs may be found in the tall grasses of meadows lying adjacent to ponds or marshes.

The leopard frog is a showy and alert-looking exhibit in the terrarium. His food requirements are easy to meet—any kind of worms, flies, or other insect life two or three times weekly.

THE PICKEREL FROG

Young pickerel and leopard frogs are the cause of that form of midsummer madness which causes middle-aged gentlemen, sedate except when carrying fishing rod and can, suddenly to drop everything and execute a series of floundering and imitative leaps in pursuit of a small and very active frog for bait.

Somewhat like the leopard frog in general size and coloration, the pickerel frog, *Rana palustris*, is often confused with it. The coloring of the pickerel frog runs to shades of brown with highlights of rich metallic bronze and gold rather than to the greenish tints of the leopard frog. Unlike the round spots of the leopard frog, with their narrow, defining rim of creamy yellow, the spots of the pickerel frog are angular, often almost square. A further difference and perhaps the most obvious one, is the bright orange-yellow shading of the undersides of the back legs and hinderparts of the pickerel. Like the leopard frog, the pickerel has two conspicuous lateral ridges running from the back of the head to the pelvic regions. The legs are cross-barred or irregularly spotted.

The pickerel frog has an advantage in life in that it is blessed with an unpleasant, somewhat acrid-smelling skin secretion which renders it inedible to any except the less discriminating animal palates. This secretion is poisonous, and many a collector has found his catch of mixed frogs killed by the presence of a few pickerel frogs unwisely tossed into the general bag.

Although usually found in the vicinity of streams and marshy ponds, the pickerel frog is far from being aquatic. From choice, it ranges the grasses and tall weeds at the water's edge in search of flies, caterpillars, moths, butterflies, gnats and similar game. However, it is to the water that it flies for safety from enemies and it is from the water in May that the male sends out the vibrant and preposterous snore which is its mating call.

The range of the pickerel frog is limited to the eastern states, but in these regions it is exceedingly common.

THE BULLFROG

The call of the bullfrog, *Rana catesbeiana*, is indeed like that of a bull, but certainly not in the least like the short, peremptory bellow of an irritated and sharp-tempered bull. Rather it is like the voice of a bull that has decided June nights are lonely and sings to relieve a slight melancholy. The bullfrog's calling is ordinarily heard from the shallows of lakes and large ponds—being our largest frog, the bullfrog naturally likes a large puddle —and is a courting song. Only the males call, and by their resonant notes attract the females to their vicinity for mating and egg-laying.

Except during the short season of mating in early summer, bullfrogs are solitary. Almost entirely aquatic in habit, they find their food—tadpoles, smaller frogs, insects and the like —in the jungles of waterside vegetation. In addition to serving as a hunting ground, these enmeshed growths give concealment and safety from the sharp beak of the heron, the underwater slash of the snapping turtle, and the gig hook of man.

Although subject to a great deal of variation, the color of the bullfrog is usually olive-green or greenish brown, mottled with darker values of the body color. The underparts are white with slightly darker spottings. The sturdy arms and powerful swimming hindlegs are irregularly spotted or barred. The throat of the male ordinarily is washed with clear yellow. Another sexual distinction is that the size of the tympanun covering the ear of a male is much larger than the eye, while in the female it is about the same size.

The back of the bullfrog is unridged by the heavy lateral folds so prominent in the green, pickerel and the leopard frogs. Instead, a characteristic fold, beginning behind the eye, loops up and around the tympanum in a half-circle and descends to the shoulder.

Since the bullfrog is solitary in nature, he had better be so in captivity. Put in with almost anything he can swallow, he is soon alone.

In an outdoor pool, the spotted and painted turtles, and those of its own species and size, are safe in the bullfrog's company. Such a pool, well planted with aquatics and with plenty of shelter, provides ideal quarters. If kept indoors, the bullfrog should have a cool tank with a bench sloping down into water deep enough for submersion. Its food may be almost anything in the way of insect or small life available. Strips of beef, wiggled in front of the frog's nose in imitation of a worm, often provoke a feeding reaction.

The natural range of the bullfrog was originally throughout that part of the United States lying east of the Rockies. In late years, however, the species has been introduced into several localities of the west coast and has multiplied there.

The larger Rana seem to run to barnyard names and noises. The pigfrog of the south, *Rana grylio,* similar to the bullfrog in size and habit, has a call very much like the grunting of a large and discontented hog.

THE GREEN FROG

The green frog of eastern North America, *Rana clamitans,* also known as the pond frog, ordinarily has head and shoulders of bright green, shading into olive or brown on the body. Like the bullfrog in general character, the green frogs are often mistaken for young bulls but may be distinguished from their larger relative by the presence of the two ridges or lateral folds down the back, which the bullfrog lacks.

The green frog is almost entirely aquatic and often frequents the shallower waters and the lower reaches of small streams emptying into bullfrog ponds.

They hibernate in the mud and vegetable debris around the edges of their water home. In regions of comparative warmth, they often remain out practically all winter. Although emerging in the very early spring, they are late breeders and it is sometimes May before their eggs are in the ponds. Unlike the quick development of the wood frog, the green frog and the bullfrog have a long tadpole stage and do not metamorphose until the second summer.

The marsh-stream terrarium, in some one of its variations, provides a natural and comfortable home for green frogs. Solitary in nature, they should not be crowded in captivity.

THE WOOD FROG

The small wood frog of the northeastern states, *Rana sylvatica,* is, after the spring peeper, the first voice of spring. In March, when the tree tops are still a welter of bare branches against a windy sky, they come from their hibernation underneath old logs and forest debris. The skunk cabbage is no more than a bundle of leaves tightly folded in a sleepy spike when their quacking note begins to sound from ice-rimmed ponds. These are the males. Female frogs and toads do not call although they often croak or scream. Vocal sacs are the property of males only. The call is made by the rapid passage of air back and forth between the lungs and the resonant throat pouches, thus producing a vibration of membranes in the larynx similar to our vocal cords. This closed circuit voice permits the frog to call from under water as well as above. Often the curious human beside a pond is surprised at hearing much frog but seeing none. In the tree toads and the toads, the vocal pouch usually balloons out under the chin and pulsates with the call. In the frogs, the pouches are similar or are evident only in a prodigious swelling of the sides of the neck.

The development of young wood frog tadpoles is rapid and, early in June, hordes of tiny frogs emerge from ponds in which the eggs of March were laid, to make their way into the deeper woods. Their coloring of reddish brown is an admirable imitation of the dead leaves littering the forest floor. A conspicuous marking is the triangular dark brown patch behind the ear, covering the tympanum.

THE WESTERN RED-LEGGED FROGS

The red-legged frog of the northwest, *Rana aurora aurora*, and its subspecies, *Rana aurora draytonii*, are somewhat similar in appearance. Young specimens are easily confused. *Aurora* is delicately formed and shy; it proves difficult to keep in the terrarium unless the temperature and humidity are just right.

Draytonii, less a creature of the dense forest, ranges through Oregon and the less arid regions of California. Semi-aquatic and hardy in captivity, it is an admirable possibility for the marsh-stream terrarium. Dark brown in color, somewhat rough-skinned, sometimes with darker spots and leg bars, its conspicuous feature, like *aurora*, is the red inner surfaces of the hind legs and the red mottling along the sides. The largest frog of the west coast, it came to the attention of San Francisco's French colony during gold rush days and quickly became a favored delicacy.

Rana pretiosa, the Pacific spotted frog, occurs throughout essentially the same range as *draytonii*, but is nowhere overly abundant. Much like the eastern green frog in size and general form, the coloring of *pretiosa*, like that of the red-legged frog whose range it shares, is in values of red and red-brown. The undersurfaces of the legs are a shade of salmon, and a streak of the same color edges the grayish brown ground color of the belly. The reddish or yellowish brown of the back is ordinarily broken by a few irregular dark spots between the lateral folds running from head to pelvis. *Pretiosa* is largely aquatic and has been much blamed for the destruction of small game fish.

THE CRICKET FROGS

The tree and cricket frogs, much smaller and more delicate than the ground-dwelling and aquatic frogs, may easily be distinguished by the presence of adhesive discs for climbing on the tip of each finger and toe. In the tiny cricket frog, *Acris gryllus,* and the swamp cricket frogs of the genus Pseudacris, the discs are present but are so small as to be hardly noticeable. These species climb but little. Their home is on the land, in the grasses and vegetation bordering streams and swamps where insect food is abundant. Although they take refuge in the water in emergencies, it is not their favored element. They are indifferent swimmers and lack large swimming webs on the hind feet such as the Rana have.

The tiny *Acris gryllus*—scarcely an inch in length—is exceedingly active and able to make high and long jumps. This ability has given it the common name of grasshopper frog in many localities.

The swamp cricket frog, *Pseudacris nigrita triseriata,* and several related species and subspecies, range from New York and New Jersey west as far as Utah and Nevada, and throughout most of the South. All are small, slender and very active frogs with pointed heads. The toe discs are very small. A conspicuous character is the extreme length of the fourth toe of the hind foot. Most Pseudacris are striped lengthwise with dark brown markings against a ground color of olive, olive-brown or gray-brown.

Being so small, the cricket frogs should not be placed in a terrarium with the cannibalistic Rana or other large amphibians but should have a tank of their own, half-water, half-land, planted with the characteristic vegetation of their natural home. Their food may be anything of a living nature small enough for them to handle, such as fruit flies, white worms and small wax and meal worms.

THE SPRING PEEPER

On warmish midwinter afternoons, when the trickling of thawing ice and snow makes a promising and pleasant little undertone in wet woodlands, there sounds an occasional tentative and piping little note, questioning and sleepy. From its winter quarters under the mosses and dead leaves, the spring peeper, *Hyla crucifer,* is beginning to prophesy, often prematurely, the imminent coming of spring.

The startling fact about the spring peeper is the discrepancy between size and voice. They are the smallest of our tree frogs, with bodies less than an inch in length; yet their shrill and assertive chorus, heard in the night, seems as if it must surely be the singing of large frogs.

The specific name, *crucifer*, is derived from the Latin and refers to the dark X-marking on the back. The color is variable and, as in practically all tree frogs, changeable according to conditions of light, temperature and humidity; but it is usually some shade of brown with washings of light yellow on the underparts.

Found through practically all of the United States and southern Canada east of the Rockies, tree frogs are available almost anywhere. Collecting at night by trailing the individual call with a flashlight is the easiest method. In the daytime, peepers frequently call during showers or when it is cloudy overhead, but at such times their voices are apt to come from the concealment of leaves or the depths of moss and grass clumps.

They are easily maintained on a diet of small insects, worms and flies. Their terrarium may be a waterside planting of sagittaria, marsh marigold, sedges and similar plants.

THE GRAY TREE FROG

During late spring in the New England and eastern states, the common tree frogs, *Hyla versicolor*, leave their winter hiding and gather to breed. For several nights the bird-like trill, sustained and resonant, sounds in the vicinity of favored ponds. Often, calling begins in the later afternoon, not from the pond itself, but from nearby trees and vegetation; later in the night, it resounds from the water's edge. A tree frog calling is intent on his music and is not greatly disturbed by a flashlight. In its beams, the throat's large pulsating sac may be seen, and, if the collector is interested in photography, flashlight photographs of the performance are comparatively easy. A visit to a pond where *versicolor* are calling often reveals the presence of males only, all trilling lustily. The silent females are not easily found but, if the time is right, are to be later seen traveling over the boulders and fallen debris of the shore to the calling males.

The eggs are attached to plant stems near the surface of the water, either singly or in groups, and hatch in two or three days. Instead of being dull and dingy in color like most tadpoles,

D. *Dwight Davis*

RED EFT STAGE OF THE EASTERN NEWT

C. *M. Bogert*

WESTERN NEWT

Robert Snedigar

TIGER SALAMANDER TAKING MEAT FROM FORCEPS

Courtesy General Biological Supply House, Inc., Chicago, Illinois

AXOLOTL

the *versicolor* are at first a light yellow in color with golden glints in the skin, orange-red eyes and a tail beautifully washed with red. The development is rapid and they metamorphose into the adult form within a couple of months.

In many localities this tree frog is called the tree toad, because of the rough texture of its skin. In size, it is large for a tree frog, measuring ordinarily about two inches in length. Its normal home is in the trees where it finds not only a hunting ground but protection. So exactly does its greenish gray or brownish gray match lichen-covered trunks or mossy bark that it defies detection. Its color is changeable to a degree. In the lighter phases a dark, irregularly star-shaped spot is seen on the back and dark cross bands bar the legs. In darker phases, these patterns are almost obliterated by the prevailing tone. The under parts of the hind legs are washed with bright orange.

The adhesive discs on the toes are well developed and very sticky. In hunting, its activity often turns into a series of frantic acrobatics. Quite commonly an insect two or three feet away is the objective of a combined flying leap and outflip of the sticky tongue. The leap is usually successful, but at the end the frog is often seen hanging by one foot, trying wildly to get a better hold.

The tree frog is most satisfactory in the terrarium, and can be kept over long periods. Its normal diet may be replaced by meal worms and earthworms. After the frog has become accustomed to handling, these are willingly consumed while the eater clings to a human thumb. In managing a large earthworm or big insect, the front feet are used to cram the obstreperous morsel into the mouth.

In nature, tree frogs are often heard calling at other times than the breeding season. As the call is popularly supposed to be a prediction of rain, this has gained for them the name of rain frogs. In captivity under good conditions, they often chirp sleepily at most unexpected times. A couple kept for two winters in my home terrarium had a habit of making a few half-trills very early almost every morning. As nearly as I could tell, their call was stimulated by the *klop-klop* of the milkman's horse in an otherwise dead-quiet street.

THE GREEN TREE FROG

In addition to the peeper and the gray tree frog, a number of other species of the genus Hyla range through various parts of the United States. Varying in size and coloring, all are similar in the possession of sticky toe-pads, balloon-like vocal pouches and largely arboreal habits.

The green tree frog of the South and the Mississippi Valley, *Hyla cinerea*, is a little longer-legged than the other North American Hyla. Its skin is less granular than that of *Hyla versicolor* and the coloring a shade of bright green, although at times it darkens and appears almost black. The undersurfaces are white or creamy white. Its home is not in trees, but in the higher swamp and stream vegetation. Professor E. A. Andrews found a number of *cinerea* snugly ensconced in the trumpets of the southern pitcher plant, *Sarracenia flava*, presumably lying in wait for flies attracted by the liquid within the cup.

Breeding of *cinerea* is normally in spring, but may occur much later, as is the case with many other frogs depending in part upon temporary pools for an egg-laying site. One July, at Garnett, South Carolina, our arrival coincided with the breaking of a persistent dry spell by heavy rains. Pools formed quickly and considerable water collected in a large pig pen on the Davis plantation, inhabited by hogs with tempers ranging from curious and mild to sleepy and mean. We were attracted to it by a large mixed chorus and we found the common tree toad, the green tree frog, the southern toad and the squirrel frog, all trying to lay delayed eggs within the limited waters. In addition, we heard a peculiar note in the chorus, almost exactly like that of a bleating lamb; and after a long hunt found the tiny narrow-mouthed toad, *Gastrophryne carolinensis*, using the puddles as well.

The green tree frog is not at all difficult to keep in the terrarium. Although willing to eat meal and wax worms, it has a fondness for flies and, if possible, these should be supplied.

THE PACIFIC TREE TOAD

The Hyla of the west coast, *Hyla regilla*, although called a tree frog, is less arboreal than most of its relatives. Like the spring peeper of the East, it is more often found hopping about in low-growing vegetation or among dead leaves of the woods floor.

Small, exceedingly active and long-legged, *regilla* is not easy to catch. Its variable color defies description. More than any other of our frogs, it has the power of quick change and may in the course of a few moments' time run from a clear light tan or green through a series of spotted and barred patterns with hints of red and gold to almost black. It is often found the bright and clear green usual in *Hyla cinerea*. In a batch, no two are alike for very long and the impression such a collection gives is that of a mixed group of related but differently colored and marked species.

Hyla regilla proves hardy in the terrarium but tends to spend most of its time concealed beneath moss and leaves and is much more likely to be heard than seen.

The salamanders are the least familiar of our small creatures and the least understood. As a rule, they are denied even a name of their own and forced to masquerade under the ambiguous and inappropriate term "lizard." The true lizards are endowed with scaly skins for protection and the conservation of moisture. The salamanders have no such covering. Their skins are mostly thin, moist and similar in respiratory function to the skin of the frog. The principal point of resemblance between the salamanders and the lizards is in the mutual possession of four legs and a long tail.

In European countries, it has been supposed for centuries that salamanders are born of fire, and this legend has in some localities been transferred to our own native species. Creatures of the dark, salamanders usually lie concealed during the day. Cracks and crevices of old lumber and logs are favored hiding places, and often they are brought into the household with wood. Assailed by the smoke and heat of their burning home, these poor habitants attempt to save themselves and, tumbling through hot ash and fire, seem indeed born of flame.

Physiologically and in life history, the majority of salamanders demonstrate their close relationship to the frogs and toads. With many species, the eggs are laid in the water and the young, after a swimming larval stage, metamorphose into the adult form and take up a terrestrial life. This, however, is not a constant pattern but has been modified in a great many forms to cope with the problems of adjustment to environmental conditions.

THE EASTERN NEWT

In the eastern and southern states, the most frequently encountered salamander is the common newt. *Diemictylus viridescens,* often called the green water "lizard." Abundant in most parts of its range, the newt has a singular beauty and delicacy of coloring, interesting habits of life and, in the aquarium, no shyness or desire for privacy. Of all salamanders, it is the toughest, the most attractive and amusing in behavior and the easiest kept and fed.

Adult newts are normally aquatic but are without gills. Their breathing is done by means of skin and lungs. They are found, sometimes in great numbers, in the quiet and vegetation-clogged waters of ponds and slow streams. Three or four inches long, green or yellow—olive above and lightly speckled buff beneath, they are distinctively marked by a regular row of black-encircled scarlet dots along the sides. Tails are flattened for swimming, and become a little ruffled in the males before the spring breeding season. Males are easily distinguished from females by the heaviness and stockiness of the hind legs and a body less inclined to rotundity and plumpness.

After a normal larval stage in the water, and metamorphosis, the newts of most localities leave their aquatic habitat and for a time live on land, sometimes fairly far from their native water. These adolescents, known as efts, are usually encountered in the woods after a rain, crawling out from under moss banks or masses of dead leaves. Clear orange-red, they are marked like the adult newts with a row of black-encircled scarlet dots along the sides. The skin of the newts, unlike that of most salamanders, is not slimy but is somewhat granular in texture. In the eft stage particularly, the skin is rough and if one of these little creatures is put in water, it takes with it a thin and silvery coating of air.

In some localities, such as Long Island, this red eft stage is not possible because of conditions unsuited to land life, and the newt larvae metamorphose directly into the swimming adult form.

In a well-planted, sunny aquarium newts breed readily. Their food in nature is all sorts of small aquatic life. In captivity, finely cut meat and liver, bits of fish and, if possible, small earthworms, white worms or tubifex, fed every other day, furnish an adequate diet. Both the red eft and the newt may be kept in a tank combining water and land areas, although the eft is much more decorative placed among the mosses and ferns of the woodland terrarium. Without white worms or very small earthworms, red efts are difficult to feed. Either they do not like meat or fail to recognize it as food because of its lack of movement. If given worms, they quickly learn which part of the terrarium is used for feeding and haunt it expectantly.

THE WESTERN NEWT

The western newt, *Taricha torosus*, also called the water dog, is much larger than its eastern relative, attaining as much as seven inches in total length. Like the eastern newt, it combines aquatic and land life but in a different fashion. After metamorphosis from the larval aquatic stage, the adolescent water dog takes up life on land as a brown, unspotted, rough-skinned salamander with buff-yellow underparts. With the return to the water at maturity, there is no great change in color, but the round tail becomes finned and adapted for swimming and the skin of the males becomes smoother. In the eastern newt, the return to the water is final and the greenish adult remains aquatic. The western water dog reserves the right to take up a land life again after having been in the water as an adult. This is a necessary adaptation to an environment in which ponds and streams of many localities do not last over a long dry season.

Like its eastern relative, the western newt eats almost anything and in the aquarium or terrarium finds a diet of earthworms, meat and fish satisfactory and adequate. An occasional dainty morsel eagerly devoured by the newt of either species is its own periodically cast-off skin.

THE MUD PUPPY

Some salamanders are entirely terrestrial and have no aquatic stage. Others come to maturity in the water, retaining their gills and swimming tails, and remain aquatic always. Species of this habit are called *neotenic*. Our best example of the retention of larval characters permanently is the mud puppy, *Necturus maculosus*, and its related species. Common in the lakes and streams of some areas of the eastern half of the United States, Necturus is sometimes found in such numbers that it is a serious problem because of its destruction of fish eggs.

Its coloring of soft brown with a mottling of blue-black on the sides and back is in agreement with the shifting light and color of the stream and lake beds over which it drifts in search of food. It dislikes light and spends most of the daylight hours jammed under or alongside rocks or logs waiting for floating edibles. The body of the mud puppy is streamlined from the rounded head to the wide swimming tail. Ineffectual little legs are hardly seen at first glance. The large external gills are the most striking feature. In cool, well-aerated water, these usually lie close to the head and are dark in color; but in stale or warm water, the gills spread out in an effort to get all available oxygen; the tiny processes become bright red, and to circulate water through them faster they take up a rhythmic motion. Skin respiration and the use of lungs are also important means of breathing for Necturus. The two together sometimes serve to take over the function when the gills have been severely damaged or lost.

The nest of the female mud puppy is a shallow cavity beneath a flat rock or submerged log. The eggs, each in its individual gelatinous capsule, are fastened to the ceiling. The number is variable, but nests have been found which contained as many as one hundred and eighty eggs. In these cases there is a possibility that the batch represented the laying of more than one female. The female remains with the eggs during incubation, but her protective ability and instincts are questionable. Cases have been noted in which it was obvious that a whole clutch of eggs had been stolen by creatures small enough to have been routed by the supposedly on-watch mother.

In the aquarium, Necturus need, if possible, cool running water of several inches' depth. Failing running water, give their tank a cool location and change the water daily. Piles of flat rock for hiding places are very necessary. Necturus are sensitive to light and in a large tank may be herded and driven from one end to the other by the beam of a flashlight.

In nature, their food is made up of worms, the eggs of fish and other salamanders, aquatic insects and other water-carried victuals. In the aquarium, they readily snap up earthworms, or strips of beef and liver. A heavy feeding twice weekly is enough.

THE HELLBENDER

Like the mud puppy, the hellbender, *Cryptobranchus alleganiensis*, spends its entire life in the water. There are no external gills as in Necturus, but gill slits opening into the throat are present as well as lungs. In spite of these respiratory mechanisms, the hellbender seems to depend mostly upon skin absorption for its supply of oxygen.

The most striking things about the hellbender are its size—it

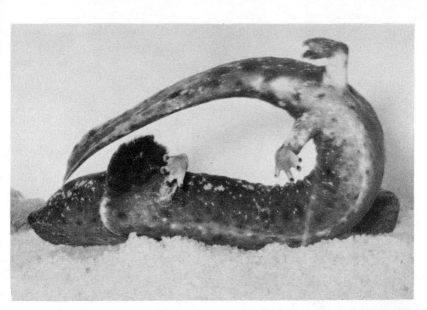

MUD PUPPY

Robert Snedigar

MARBLED SALAMANDER WITH EGGS

John C. Orth

D. Dwight Davis

Marbled salamander

D. Dwight Davis

Red-backed salamander

attains as much as eighteen inches in length—and its extraordinary homeliness. It is almost beyond belief that a living creature could look so unpleasantly like a piece of a very old and water-logged boot. The eyes are small and almost invisible in a wide flat forehead. The flabby, compressed body is edged with folds of loose and slimy skin. A dweller among rocks and roots, the hellbender is protectively colored a dull and muddy olive or reddish brown. Its repulsive aspect has given it a bad reputation throughout its range in the Mississippi Valley and the central states, where it is popularly believed to be venomous. Actually it is harmless.

The hellbender nest, a basin-shaped cavity scooped in the stream or lake bed, is made and guarded by the male. After depositing the long bead-like strings of large eggs, the female is chased from the scene and not permitted to share in parental care or glory. The male guards the eggs with a haphazard zeal and very often, absent-mindedly perhaps, gobbles up a few himself.

In the aquarium, *Cryptobranchus* needs essentially the same conditions and food as the mud puppy.

THE TIGER SALAMANDER AND THE AXOLOTL

One of our most widely distributed North American salamanders combines neoteny and a terrestrial life. The tiger salamander, *Ambystoma tigrinum*, follows a very conventional pattern in its normal life history. A migration of males and females to the breeding ponds occurs in early spring. For a day or so, the selected water is the scene of courtship and mating. In salamanders, eggs are not fertilized as they leave the body of the female. The males deposit on the pond bottom small gelatinous lumps tipped with a whitish ball containing hundreds of spermatozoa. The ripe female crawls over a spermatophore and, in passing, takes the fertilizing tip within her cloacal chamber. Liberated there, the spermatozoa are within easy access of the unfertilized eggs. The eggs, each enclosed in a gelatinous envelope, are laid within the next few hours, and hatch in a few weeks' time.

The larvae are small and seem to be mostly head and bushy growths of gills. By fall, they have attained considerable growth and are ready to metamorphose and take up land life. The gills shrink and are absorbed as lungs develop; the tail loses its swimming contours. Soon we have a blunt-headed, blackish salamander, rather chunky and heavy-looking, with mottlings of dingy olive-brown over head and body. On emergence from the water, the young tiger takes refuge beneath woods debris or in burrows and is not likely to be seen again until its own breeding season comes.

The foregoing is the normal development of *Ambystoma tigrinum*. In the highland lakes of Mexico and the Rockies, it is different. There the tiger salamanders never normally metamorphose but grow to adult size as gill-retaining, lungless, swimming creatures. This neotenic form of the tiger salamander was first found in Old Mexico by the Conquistadores and still retains the Aztec name of axolotl. Although larval in other characters, the axototl becomes sexually mature and, in cold lakes, generation follows aquatic generation. The chain may be broken. Low water or other circumstance unfavorable to a continued water life may bring on metamorphosis into the terrestrial tiger salamander form.

Axolotls are not uncommon in the market and even an albino form is frequently to be had. In keeping them, fairly deep and cool water is necessary. If conditions are not exactly to their liking, metamorphosis may ensue.

THE MARBLED SALAMANDER

The genus Ambystoma is perhaps our most widely distributed group and, in addition to the tiger salamander, contains a number of species more or less commonly encountered.

The marbled salamander of the eastern and central states, *Ambystoma opacum,* is smaller than *tigrinum* and more delicately formed. Black, with fairly regular markings of an almost pure white, *opacum* may easily be distinguished from its relatives. Breeding habits vary from those of the rest of the group in that the eggs are laid on land in the fall under rotton wood and leaves. The female remains on guard until they hatch during the rains of spring and the gilled larvae can find a watery path down to the ponds.

THE SPOTTED SALAMANDER

A pattern of scattered round yellow spots on a dark bluish or black ground distinguishes the spotted salamander of the eastern and central states, *Ambystoma maculatum.* Almost as large as the tiger salamander, the spotted is one of our most handsome forms. Its usual hiding place is in the woods beneath leaves and trash or in burrows, but it is occasionally found crawling about in the dampness of farmhouse cellars.

THE LONG-TOED SALAMANDER

In the northwest, a particularly attractive little Ambystoma is found. Called *macrodactylum*, because of the length of its toes, it lives in damp and heavily wooded areas, usually in the vicinity of cold mountain streams or lakes. Small, blue-black with a distinct yellow-green stripe down the back, *macrodactylum* is only hardy if kept in a very cool and damp woodland terrarium.

THE WESTERN GIANT SALAMANDER

Closely allied to the Ambystoma is the giant salamander of the west coast, *Dicamptodon ensatus*. This, the largest land salamander in the world, grows to a length of as much as twelve inches, and is a resident of damp coastal forest areas from British Columbia to southern California. It is shy and secretive in habit, and comes out from its hiding places only at night to hunt for worms, insects, smaller salamanders and other food. One observer found a giant salamander eating a white-footed mouse.

The breeding habits of Dicamptodon are not definitely known but are probably like those of the Ambystoma. Several cases have been noted in which larvae of Dicamptodon became mature without complete metamorphosis, producing a form similar to the axolotl.

The Ambystoma and closely related forms have lungs in the adult stage. Another group, which includes the majority of our North American salamanders, is lungless and depends for respiration upon the skin and the highly vascularized membranes of the mouth and throat. This lungless family, the Plethodontidae, includes aquatic, semi-aquatic and terrestrial members. A large number of plethodontids occur in North America, many of which are localized and have but a limited range. All of them are shy and secretive; very few are to be classed as common or frequently seen even in the regions of their greatest abundance.

THE RED SALAMANDER

Of the aquatic plethodontids, the red salamander of the eastern states, *Pseudotriton ruber,* is the most attractive. The dark coral-red adults, patterned with an irregular spotting of black, are found in and around springs. The eggs, each in its individual gelatinous capsule, are laid on the under-surfaces of stones in deep, cold waters. A characteristic of the plethodontid egg is that each is attached by a slender stalk. The larvae of Pseudotriton are lacking in the brilliant color of the adults, but begin to take on a reddish tinge as they approach metamorphosis.

The genus Eurycea of the eastern states is closely related to Pseudotriton. The long-tailed salamander, *Eurycea longicauda*, and the cave salamander, *Eurycea lucifuga*, both have a light salmon-orange coloring with black markings which might be mistaken for the coral-red and black pattern of Pseudotriton. The slimmer body and tail and large eyes of the Eurycea should serve to distinguish them. In addition, Pseudotriton are found almost always in the water; the Eurycea are more terrestrial and are found in the debris along streams. The cave salamander has frequently been found about the openings of mountain caves.

THE DUSKY SALAMANDER

Throughout the eastern states the dusky salamander, *Desmognathus fuscus*, and its subspecies are more often found than any of its close relatives. Very active, slippery and hard to catch, the adult duskies hide underneath the logs and stones lying along the edges of brooks or in the moss and damp debris nearby. Frantically, and with a characteristic wiggle, they rush for the safety of deep water or a convenient crack or crevice when uncovered.

The breeding habit of the dusky is of particular interest. Although a semi-aquatic creature, the dusky's eggs, clustered in grapelike bunches, are laid on land in a hole dug under leaves or debris. The nest is dug by the female and is large enough to accommodate not only the eggs, but her as well. She lies in such a position that her twisted body contacts practically the whole mass. Unlike most salamanders, the young, at hatching, are not gilled larvae, ready to take to the water, but must spend two weeks or more as tiny land-dwellers before they develop gills and are ready for a normal amphibian larval life.

The dusky salamander is medium in size and, as its popular name indicates, not brilliantly colored. A soft brown or dark brown is the usual tone, relieved in many cases by markings of light reddish-brown along the back.

RED SALAMANDER

LONG-TAILED SALAMANDER

SLIMY SALAMANDER   *D. Dwight Davis*

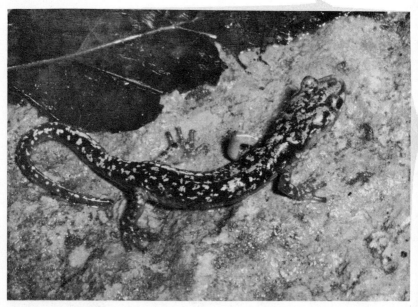

D. Dwight Davis

EASTERN COUSIN OF WESTERN TREE SALAMANDER

In addition to *fuscus*, several other species of Desmognathus have limited ranges in the eastern states. Some of them are larger in size, but all have much the same character and living habits.

THE RED-BACKED SALAMANDER

Another common species of the eastern states is the terrestrial red-backed salamander, *Plethodon cinereus.* Small and slender, the red-back is marked with a straight-edged band of reddish-gray or gray along the back and tail. Like the dusky mother, the female red-back remains on guard over her eggs. Instead of hatching out larval salamanders of aquatic habit, these red-back eggs bring forth individuals that have already undergone their metamorphosis and are miniature adults. Instead of scattering, the young remain with the mother for a time. Her moist body helps to keep them from drying out and perishing in the dry fall woods.

THE SLIMY SALAMANDER

The slimy salamander, *Plethodon glutinosus,* is large, growing to as much as seven inches in length, almost black, lightly dotted and blotched with silvery white along the sides. It is fairly common throughout the eastern half of the United States. Its skin is conspicuously provided with mucus glands and the copious flow of their secretion whenever the salamander is handled is good reason for the popular name. The slimy and other related species of Plethodon are found on land, usually in forested areas under logs and other woods litter.

THE WESTERN TREE SALAMANDER

In the West, two interesting genera of lungless salamanders are found. Aneides, in several species, lives on the ground underneath boards, in the crevices of rotten logs and also in the hollows of trees. Cavities in the California live oak are a favored hiding and nesting place. The habits of egg-laying and guarding are similar to those of eastern plethodontid salamanders.

The Aneides have prehensile tails and commonly pose with the body in a twist and the tail coiled to the side. *Aneides lugubris,* perhaps the most common species, is a light, transparent tan salamander with creamy under parts. Specklings of the lighter coloring mottle the sides, legs and tail.

BATRACHOSEPS

The California salamander, *Batrachoseps attenuatus*, and its several subspecies, are unmistakable. The possession of a tail twice the length of the slender body and head, short legs and quick wriggling movements make them almost wormlike in character. They are dark brown or blackish in color—much the tone of dead leaves—and are commonly found under rotten wood and bark. Although they stay above ground all year in the more humid parts of their range, Batrachoseps disappear in the dry regions with the approach of summer, and sleep in the security of some deep earth crevice until the fall rains again make life above ground possible.

The matter of housing amphibians is neither difficult nor expensive. The smaller frogs and toads are best cared for in screen-covered fish tanks or in large jars. For those of aquatic or semi-aquatic habit, the marsh-stream terrarium is well suited. Tree toads are climbers, and their container should be high enough to hold an upright heavy-stemmed plant or a piece of tree limb. Toads of a burrowing habit must be given enough earth to cover themselves.

In housing salamanders, the habits of the species must also be considered. If a dweller along the borders of streams, water and land areas may be combined in a terrarium. If terrestrial, all that is needed is an adaptation of the woodland setting.

The great difficulty in maintaining amphibians is to keep them cool enough. In very warm localities or indoors in the winter, particularly if animals are kept segregated in individual dishes or jars for experimental work, a flow of water around the containers is often necessary to keep the temperature down.

High temperatures encourage the development of fungus growths which attack both the skin and the intestinal tract of amphibians. These molds most commonly affect frogs and aquatic and semi-aquatic salamanders. They are easily recognized by the characteristic patches of whitish fuzz on the affected parts. The disease is very infectious and spreads quickly unless checked by the elimination of sick animals, careful and clean handling and proper sterilization of containers and implements. The disease is almost always fatal, although if the condition is noted early enough, a five-minute bath in a solution of potassium permanganate the color of red wine, followed by a week or so at a temperature of around 50°F., may check it. An old-fashioned ice-box is good for such hospital cases. Unless the affected frogs or salamanders are not replaceable or are a part of an important study series, this treatment is hardly worth the bother and the animals are best sacrificed for the common good. Whatever the disposition of the infected creature, its tank should be sterilized before receiving another inmate.

These fungus diseases are not peculiar to captive amphibians. Newts in ponds sometimes die in great numbers before cold weather checks the spread of an infection.

Aquatic frogs are frequent victims of a bacterial disease known as "red leg." This highly infectious complaint commonly appears first in a specimen weakened by injury. Once well started, it spreads rapidly and may destroy the entire stock. The major symptom is a red flush over the underparts, especially of the hind legs. This is caused by the congestion and breaking of small blood vessels just beneath the skin. The kidneys are affected and are unable to eliminate the moisture absorbed by the skin, so that if the frog is kept in water, the body tissues become swollen and saturated. Recovery from red leg is doubtful, but may be materially assisted by keeping the patient on damp—not wet—sterilized moss at a temperature of between 40° and 50° Fahrenheit.

With a few exceptions, all amphibians are carnivorous in the adult stage and for their feeding we are dependent largely upon supplies of earthworms, meal worms, crickets, grasshoppers, flies, white worms and other small living creatures which may be available or which can be raised. Large frogs gobble up smaller frogs. For the most part, large salamanders are also cannibalistic and make a meal of smaller species. Of the entire group, the only ones which may be consistently fed bits of meat and liver are the aquatic salamanders. Frogs, toads and salamanders of terrestrial habit recognize prey only if it moves. In exceptional cases, skillful teasing with a bit of meat impaled on a toothpick produces a feeding reaction in one of these problem boarders. For most grown amphibians, a feeding twice weekly is sufficient, although some species, such as the bullfrog, are greedy and always willing to stuff. The cheap spade-shaped forceps made for stamp collectors are most useful for feeding not only salamanders, but many small creatures.

Frog and toad tadpoles are largely vegetarian feeders. Salamander larvae are carnivorous and when very young need such small live foods as daphnia, tubifex or white worms. In a group of young tiger salamander larvae, certain individuals will be seen to take on growth with extraordinary speed, and in a short period becomes many times the size of brothers and sisters from the same batch of eggs. Cannibal is a harsh word to apply to creatures so young. Let us say that if the growing individual be left with his fellows, in due time the entire brood will be merged into *one* large individual. Encouraging such a merger is risky if several of the youngsters have the same idea. They have but little size discrimination and the poetic justice of choking to death on a brother too large to swallow is often visited upon them.

Lizards

SUPERFICIALLY LIKE the common salamanders in form and appearance, the lizards are an entirely distinct group, related to the snakes, crocodiles and turtles, and with them forming the great class of the Reptiles. The external character which sets the lizards unmistakably apart from the salamanders is the texture of skin. Instead of a soft and slimy covering with many glands, the lizards have a skin made of horny scales, glandless and dry to the touch. In behavior and living habit, the salamanders and the lizards differ widely. The first are clumsy and awkward on land and seek escape in their native water or by burrowing beneath stones or leaves. Lizards, almost without exception, are quick and agile and lead the would-be captor a merry chase. Terrestrial salamanders are creatures of the dark and damp, with no love for the sun; lizards for the most part are animals of dry and warm places, unable to live long without the sun's health-giving rays.

The arid regions of the West and Southwest are rich in many forms of reptilian life. Lizards of many species abound. Their varied sizes, colorings and habits of movement and hunting offer a miniature picture—very small indeed—of what life in this world may have been in the long ago Age of Reptiles when giant lizardlike creatures possessed the earth. Like old Tyrannosaurus, the leopard lizard rears upon hind legs and harries the smaller life of its territory. The modern horned toad, like Triceratops underslung and low to the ground, resembles it also in the possession of a horned and formidable head. The chuck-walla, heavy-bodied and clumsy, like old Brontosaurus which it somewhat resembles, is content to feed upon herbs and vegetation.

The parallels are many. However, our own lizards are *not* to be judged direct line descendants, shrunken and degenerate, of these one-time lords of the earth, but as remote descendants of the same original stock whose development along smaller lines has enabled them to survive in a world to which their monstrous cousins could not adapt themselves.

The largest family of American lizards is that which takes its name from the large tropical iguanas. The Iguanidae, as the family is known, includes not only such members as the chuck-walla, which slightly resembles the larger iguana in appearance but also forms which do not resemble the true iguana at all.

THE AMERICAN CHAMELEON

One of these is the small lizard known as the American chameleon, *Anolis carolinensis*. Anolis is not near kin to the true African chameleon, but like it undergoes striking color changes under the influence of fear, rage and such physical factors as changes in light, temperature and humidity. Many lizards have this trait in a less marked degree. The skin of Anolis is layered beneath with dark cells called melanophores. In a change of color the degree and character of the change is due principally to the spreading or contraction of the pigment granules in these cells and the resultant effect on light absorption and reflection.

Although there may be some correspondence between the color of the surface on which the lizard rests and its own color, such a correspondence is very doubtful. Leaf green and brown Anolis are found impartially upon green leaves and dark tree trunks. Often, instead of retaining its protective hue, a brown Anolis on wood bark may be seen to trade its inconspicuousness for a bright and showy green, when excited by the approach of another lizard or other stimulus.

Another distinctive characteristic of Anolis is the possession of a throat fan of clear reddish-orange which, by means of a crescent-shaped bone and cartilage, may be extended for display. The use of this decorative appendage seems to be a warning signal to other Anolis that the territory is pre-empted and that further invasion of it means fight.

In shape, these small lizards, with their long, pointed snouts and long tails, resemble miniature alligators.

In their preferences of humidity and temperature the Anolis are like their larger tropical relatives. Ranging through warm regions of the southeastern states and those bordering the Gulf of Mexico, they are found in great numbers where vegetation is lush and green and festoons of Spanish moss and vines offer adequate protection and an aerial hunting ground. Their frequent presence along the brushy fence lines bordering truck gardens indicates that the species is very likely of considerable economic value in insect control.

The Anolis are largely arboreal and by means of flat, adhesive discs under the toes, in addition to sharp claws, are able to swarm upon almost any surface. In captivity, they clamber over and under all cage furnishing and may often be seen hanging upside down from the screen cage top, stalking the flies buzzing there or climbing up the slick glass sides, held only by the suction of the toe discs and belly friction.

For food, Anolis are confined to small moving creatures; flies, beetles and their larvae, spiders, moths and similar insects. This makes it necessary, if Anolis are to be maintained in health, to raise or trap one or more forms of live insect food. Meal worms are acceptable, but their horny outer coat makes them less digestible than flies, moths and soft-bodied grubs such as the wax worms. If possible, the diet should be varied to include all available insect life.

Two other elements enter into the successful maintenance of Anolis. Sunshine is essential; without it these lizards refuse to feed, droop and soon die. Adequate and suitable plant material in the cages affords the inmates not only protection from imagined external dangers but permits escape from the assaults of one another. Equally important, the growing plants help keep the humidity of the cage at just the right degree. In the American Museum laboratories, large cages, profusely planted with a variety of trailing vines and semi-tropical vegetation, receiving the sun for several hours each day, house hundreds of these tender lizards. Indeed, under these conditions some of the older males have attained sizes much larger than that of field specimens.

If it is at all possible, arrange for a continuous dripping of water into a pile of rocks at one end of the Anolis cage. Otherwise, sprinkle the cage morning and late afternoon with a coarse spray so as to leave drops of water on the vegetation. Anolis seldom drink from vessels, but eagerly lap up drops of imitation dew and rain.

In behavior Anolis are spry and agile and, in a cage housing a number of them satisfactorily, something is always going on. There is keen competition for territory, and the spectacle of two males carrying on a miniature battle of the monsters—the red of their spread throat fans vivid against their emerald green and the more sober coloring of the surroundings—is but one of the dramas enacted daily.

The information desks of zoos and museums have their own way of knowing when the circus comes to town. Callers, in person and on the telephone, start a stream of inquiries about "chameleons," their habits and foods. Hawkers, with Anolis by the hundred, haunt the circus grounds and, with the assurance that nothing less spiritual than sugar water is needed to keep alive their delicate wares, persuade a well-intentioned public to buy them.

THE HORNED TOADS

Like the Anolis, the so-called "horned toads" are hawked about the streets on gala days and found in all kinds of pet shops. Of late there has been an increasing tendency to send them through the mail as tourist souvenirs and advertising novelties. The seller usually knows as little about the reptile's habits and needs as the buyer and, thus damned from the start by misinformation, the horned toad soon winds up a short captive career.

The popular name horned toad is of long standing for the iguanid lizards of the genus Phrynosoma. Misnomer than it is, the term should only be used with a clear realization that these creatures are not toads but lizards and, except for their inoffensiveness and homeliness, very little like the true toads.

Many species of Phrynosoma are found within the United States. All are dwellers in regions of little rain. The distinctive characteristics of the Phrynosoma are the flat and almost circular body, the back covered with granular scales interspersed at regular intervals with short, heavy, spinelike projections, and the more or less complete crown of flat spines projecting from the back of the head. Unlike the many lizards with more tail than body, Phrynosoma is blessed with but a short and somewhat ineffectual-looking caudal appendage.

The ornateness and size of the crown of spines on the head is subject to a great deal of variation within the genus. Some species such as *solare*, of southern Arizona and Mexico, have a formidable and almost complete circlet of horns. In other species the spines are hardly more than tubercles on the back of the head. Where the horns are well developed, as in some species, they undoubtedly serve to protect the individual. Lizard-eating snakes and animals must find them a decided inconvenience, if not an unsurmountable obstacle, in swallowing.

The spines are not the only defense of the Phrynosoma. Bluff is one of their resources. Threatened by a supposed enemy, they raise their bodies as high as short legs will allow, inflate themselves with air and, open-mouthed, give vent to a short, sharp hiss. If further provoked, little furious rushes and jumps in the direction of the aggressor carry on the bluff. In the last extremity, Phrynosoma in nature gives an exhibition of an unusual mechanism, peculiar to this group, so far as is known. A fine spray of blood is ejected with considerable force from the lizard's eye; often it spurts as far as a couple of feet. A rapid rise of the lizard's blood pressure and the subsequent rupture of certain fragile-walled blood vessels in the corner of the eye is the means by which this prodigy is accomplished. Whether this phenomenon is a defense trick designed to frighten off aggression is problematical. It is one of the unfailing sources of amusement for boys in regions where these reptiles are abundant, and no opportunity to provoke its demonstration is lost. Curiously enough, when the general superstitious attitude toward reptiles is considered, the notion that the horned toad's harmless blood might be poisonous seems never to have cropped up. In my own experience, the only reason given us for not teasing horned toads was that the fine misting of blood was no fit decoration for the whiteness of a Sunday shirt.

The Phrynosoma are, of all reptiles, creatures of the sun. It is in the hottest and driest of the Arizona desert areas that they abound and thrive. Most reptiles of these regions hole up during the heat at midday. But in the spring at that time the horned toad is sometimes encountered, skittering about in short dashes, tail twitching from side to side like that of an angry cat, stalking some insect prey. Just before sunset and the quick drop in temperature accompanying the desert night, the horned toad digs into the warm sand and sleeps out the night.

In the matter of bearing young, the horned toads are inconsistent. Some species lay eggs; others bear live young. Still others lay eggs in some parts of their range, and bear live young in nearby regions.

In nature, the food of the horned toad is all kinds of moving insect life. Like the other iguanid lizards, Phrynosoma have to depend largely on their sight for food information. In captivity, unless sunshine and warmth are adequate, they refuse to feed at all and, after a period of gradual emaciation, perish. In a cage bottomed with several inches of dry sand, kept close to 80°F. and flooded with sunshine for several hours daily, the horned lizards thrive upon a diet similar to that of other captive lizards—meal worms, fly maggots, ants, wax worms, roaches, flies and the like. Captive horned toads seem to lack decision. Often, practically standing over an insect morsel, head cocked on one side to keep track of it, the prey has crawled out of range before the lizard can make up its mind to nip. Forced feeding is out of the question if a large group of lizards is to be taken care of. If the number is small it is well to insure the taking of a good meal every other day by prying open the jaws with a manicure stick and ramming in a worm or, if insect food is short, a small piece of liver. The horned toad w.ll protest and attempt to reject the morsel, but sooner or later, if the feeder persists, it will swallow it in a fit of submission or absent-mindedness.

THE FENCE LIZARDS

Among the more primitive southern people all sorts of superstitious charms based on weird and primitive reasoning are in general use. One of these is the tying of a packet, containing the dried tail of one of the several species of lizard known to these country folk as "swifts," to the leg of a young child. Through the old belief that like brings about like, the tail of the quick lizard is intended to bring agility to the baby and to make him, in later life, light on his feet—swift.

The term "swift" is most often applied to the small lizards of the genus Sceloporus and the allied genus Uta. By far the most widely distributed are the fence lizards. These are all similar in food habits, behavior and appearance. They have for years been divided into a number of species, each with its regional races, but students of a "lumping" tendency today consider most of them as one species, *Sceloporus undulatus*. There is much to be said for this reclassification, especially from the angle of the relatively unscientific naturalist. The differences between the several groups are hardly radical and obvious enough to warrant the former division.

In all except the higher mountain regions and the more northerly parts of the country, some one of the fence lizards is likely to be found. Their distribution is not always continuous, but may be spotty and interrupted by desert or other areas unsuited to their needs. Small and agile, covered with coarsely ridged scales, each tipped with a tiny projecting spine, they are the most familiar of all our lizards. The sandy bar-

AMERICAN CHAMELEON

HORNED TOAD

John C. Orth

FENCE LIZARD

C. M. Bogert

CATCHING A LARGE FENCE LIZARD

rens of New Jersey, with their scrubby growth of jack pine, the turpentine woods of the South, and the rail fences of Illinois and Missouri give shelter to the widespread eastern form, *Sceloporus undulatus undulatus*. In the lower Rockies and great plains, *consobrinus,* the prairie spiny lizard, small, striped with a fairly distinct yellowish line on each side, is found. The forms known as *bi-seriatus* and *elongatus* are to be found between the Rockies and the Sierra Nevada Mountains, in the Great Basin. Along the west coast, from Brit__h Columbia south into the San Joaquin Valley of California, *occidentalis,* very similar to the eastern form but lighter, is the most common of the fence lizards.

These small and agile lizards are often to be seen in the morning hours perched on a fence post or lying stretched on the warm surface of a rail, soaking in the enlivening sun. Woodpiles and fallen trees or trees dead and spare of shade are favorite roosting spots. On these surfaces, the Sceloporus is inconspicuous and not easily seen. If approached, the fence lizard does not ordinarily take incontinent and sudden flight. Instead, its tactics are to slip sideways around the tree trunk or fence rail, keeping the bulk between itself and danger. Pursuit often involves half a dozen complete circlings before the would-be captor gets a chance to try a grab. Catching by hand is necessarily largely guesswork. Lizard tails are fragile and more often than not the collector's wild grab rewards him with only a wriggling scrap of that appendage.

The fence lizard's habit of permitting a fairly close approach before skittering away makes it possible to use a noose of horsehair or fine black linen thread for capture. The noose should be attached to a slender stick several feet long and slowly and without abrupt movement maneuvered over the reptile's head and tightened with a quick jerk. The lizard's interest in the collector—his sharp eyes miss no movement—prevents his giving the noose the attention he might.

Although not resident in particularly arid areas, the activity of fence lizards is dependent to a large degree upon the amount of sun received. In cold and damp weather, they lie hidden and lethargic under bark and other cover. They emerge with the sun and from its warmth draw liveliness and activity. The hunt for food begins. Rotten wood and bark are favored haunts of beetles and their larvae. Woodpiles offer quick shelter for emergency as well as a good food supply and are, in Sceloporus areas, usually the permanent home of one or more. Ordinarily the pile will be the territory of a single male and such females as stray into it. Other males, although sometimes permitted to stay, are usually not tolerated. With a great show of ferocity and a display of the bright blue coloring of his sides, the resident male blusters up to the intruder and invites him to leave. As in the case of other creatures possessed of an exaggerated sense of territorial right, the invading lizard is the victim of a feeling of inferiority which puts him at a disadvantage.

Female fence lizards are usually to be distinguished from males by the absence of the bright enamel-blue markings along the sides and on the throat. Western Sceloporus are usually called "blue-bellies" from this characteristic of the males.

Very like the *undulatus* group in general character but a little smaller in size are the several subspecies of the sagebrush lizard, *Sceloporus graciosus,* from the Great Basin and coastal areas.

THE LARGER SWIFTS

In addition to the smaller species, the genus Sceloporus includes several large and showy forms commonly called spiny swifts or spiny desert lizards. The scales of these larger relatives are heavily keeled and the ridge prolonged into a sharp, projecting spine. The natural home and origin of the spiny swifts is Mexico and there a great number of species occur in abundance. Several species, similar in size and behavior, range over the border into the United States. *Sceloporus magister,* of the Southwest, is somewhat representative of the group. Living in hot and arid areas, *magister* grows to be as much as eight or nine inches in length. Males are inconspicuously but beautifully colored and marked with slightly iridescent tones of yellow, blue and green. These tints are variable and their intensity depends apparently on the reptile's physical condition and the amount of sun it receives. The neck is almost encircled by a marking of black.

In captivity, with warmth and adequate live insect food—flies, meal worms, roaches, grasshoppers, crickets, etc.—these larger swifts are more hardy than their smaller cousins. Sunshine is essential. Without it they are sluggish and refuse to feed, their coloring grows dull and there is no activity. In sunshine, appetites are good, behavior lively, and color markings, previously unseen, take on brightness.

THE UTA

The genus Uta, unlike Sceloporus, which it resembles to a marked degree, has no such wide distribution. The several species are confined to the warm and arid Southwest. The Uta may be distinguished from the fence lizards by the less rough scaling and the fact that except on the tail, the scales have no projecting spines.

THE COLLARED LIZARD

Although limited to the desert areas of the West and Southwest in nature, a number of species of iguanid lizard are so personable and attractive in habit that they are worthwhile additions to any collection. Occasionally specimens of these may be seen in pet shops or they may be obtained from collectors in the regions in which they occur.

Among these interesting and worth-having lizards are those of the genus Crotaphytus. The most widespread of this group is the collared lizard, *Crotaphytus collaris,* so called because of the black collarlike markings on the neck. In body, these moderate-sized lizards are heavy and short. The head is large, blunt and distinct from the neck. The stout and powerful hind legs and relatively short forelimbs are evidence of its habit of rearing and running kangaroo-fashion. The colors of the collared lizard are delicate and admirably adapted for purposes of concealment in the dusty grays and greens of these lizards' arid habitat. In nature collared lizards are most inconspicuous and usually call themselves to the traveler's attention by springing up out of nowhere and zigzagging at high speed from bush to bush and out of sight. While largely ground lizards, in captivity they have a marked predilection for climbing and tend to roost habitually on cacti, rock piles, sticks or other elevations to enjoy the sun.

Strong jaws and sturdy teeth enable them to cope with more than mere insect fare. Lizards, including the spiny fence lizards and young horned toads, young snakes, grasshoppers, crickets, beetles and their larvae and an occasional helping of desert vegetation, are on the collared lizard diet list.

In captivity, unless kept in a dry, sunny and well-ventilated cage, they show little of their native appetite and often refuse to feed. Proper housing stirs up their interest in food, and meal worms, crickets, large roaches, and wax worms provide adequate fare. The collared lizard gullet is large, and without straining, one of them can engulf a mouthful beyond the ability of other lizards of similar size.

Collared lizards are wild at first, but after becoming a little used to human society, they quiet down and cease battering against the cage side whenever anyone comes near.

The closely related leopard lizard, *Crotaphytus wislizenii*, is similar in build and behavior but more slender. As might be suspected from the name, it has a pattern of rounded spots. The leopard lizard is distributed through the Great Basin area and the Southwest but is nowhere common. Similar in habit to the collared lizard, it is reputed to be more savage in its hunting and is a notorious cannibal.

THE ZEBRA-TAILED AND SPOTTED LIZARDS

The deserts of the Southwest are also the home of the several species of iguanid lizards of the genus Callisaurus. These are popularly known as zebra-tails from the sharp black bars on the under-side of the light tail. After making a quick run, the Callisaurus stop and curl the tail over the back, exposing the black bars to view.

These lizards are sand dwellers and feed on both vegetable and insect material. The larvae of beetles, grasshoppers and the like are favored prey.

The Texas species of the genus Holbrookia, *texana*, shares the name of zebra-tail with Callisaurus, and like it curls the tail over the back to show zebra markings. The Holbrookia, although similar in size and general coloring to the Callisaurus, may be distinguished from the latter by the total absence of an external ear opening.

In addition to *texana*, several other species of Holbrookia are found in the southwestern country and are called generally by the ambiguous term, spotted lizards. Small and agile, they frequent hot and arid flats and live largely upon beetles and their larvae.

THE CRESTED DESERT LIZARD

The larger iguanid lizards are found principally in tropical regions and but one species, *Dipsosaurus dorsalis*, the crested lizard or, as it is also known, the pigmy iguana, extends its range above the Mexican border into the southwestern deserts.

As compared with our other lizards, the pigmy iguana is large; but as compared with some of its huge tropical cousins, it is indeed small. Including a tail almost twice the body length, mature *Dipsosaurus dorsalis* measure about a foot in length. The body is comparatively heavy and the head small. Along the back is seen the row of enlarged scales politely called the crest. The scales of the tail are keeled and arranged in circular series while those of the body are granular. Like most lizards, Dipsosaurus has a fragile tail, easily and willingly sacrificed to avoid disaster.

In color Dipsosaurus are subject to considerable variation but are generally some shade of whitish gray or gray-brown with a net-like pattern of wavy lines enclosing lighter spots. The tail is barred above with ringlike markings.

These lizards, while insect eaters, are largely feeders on vegetation. Their marked preference for the yellow desert flowers indicates, perhaps, that sight and the recognition of color play an important part in their food-getting.

In captivity, in addition to assorted insect food—meal worms, roaches, crickets, etc.—they welcome such easily obtained yellow flowers as the dandelion, the calliopsis and coreopsis.

THE CHUCK-WALLA

The chuck-walla, *Sauromalus obesus,* like the crested lizard, is largely vegetarian, and has the same predilection for the yellow blossoms of desert plants. With the exception of the Gila monster, the chuck-walla is our largest lizard. Big-bodied and clumsy-looking, it lives in the rocky desert hills of the Southwest and, in the crannies and crevices of its home, finds quick escape from capture. In exactly the same fashion as the common toad inflates itself, the chuck-walla puffs out its already corpulent body with air. Wedged in a tight rock crevice and swollen in this fashion, he practically defies all ordinary methods to pull him out. The Indians eat this lizard and their method of extricating one so caught is to puncture the inflated body with a sharp wire and withdraw the creature.

The chuck-walla goes through a slow color change running from almost black to a light reddish gray. The change is most marked on the body and tail while the head seems always to remain fairly black.

Unfortunately, these attractive lizards are not especially hardy in captivity. Certainly sunshine and dry heat are among their great needs. Their appetites are apt to be very poor even under the best of conditions and, like many lizards, they cannot be depended upon to feed themselves. Most iguanid lizards condition readily to hand-feeding and, held lightly, a little tap on the nose with a forceps is all that is needed to make the mouth fly open. Whether the mouth flies open to receive food or to register a protest, I have never been able to decide. The important thing is that it opens and food may be dropped in. For the chuck-walla, feed bananas and cooked rice in this way, perhaps even prepared barley or rice dog food. In addition, keep in the cage, for them to browse on if they will, blossoms and stems of dandelion and other yellow flowers or, failing these, lettuce.

THE SKINKS

The American lizards of the family Scincidae and the genus Eumeces have a peculiar and distinctive texture of skin which renders them easily recognizable. Unlike the coarse and spiny scales of the fence lizards or the fine-grained skin of the Anolis, the scales of the skinks form a smooth and highly polished surface with all the beauty and clarity of color of fine enamel work.

Their legs are small and weak and not fitted to support the weight of the body high above the ground in running. This might indicate that they are slow and easy to catch. On the contrary, they are most agile and able to slither their bodies over the ground and into cover with great facility.

While abroad in the day hours and fond of sunning themselves, skinks are shy and secretive and seldom stray far from their shelter of rotten logs, piles of leaf mold or creviced stone. In a region in which they are at all abundant, a systematic search for them during the cold hours of early morning will yield some result. However, a skink found is by no means a skink caught. Even more than other lizards, they have the trick of leaving but a wriggling tail in the would-be captor's hand.

In nature, they feed upon insects of all kinds—especially the grubs and beetles which infest rotten wood—and devour, in addition, the eggs of other lizards, baby mice and similar small fare. Unlike the iguanid lizards, skinks are not largely dependent upon movement to show them food, but are able to detect still or buried edibles.

The iguanid lizards have a short and broad tongue with which they catch insects after the manner of a frog. Other

SWIFT

CRESTED DESERT LIZARD

U.S. Fish and Wildlife Service, photo by E. P. Haddon

COLLARED LIZARD

U.S. Fish and Wildlife Service, photo by E. P. Haddon

CHUCK-WALLA

families of lizards, including the skinks, have forked or notched tongues which are thrust in and out as the reptile moves or its attention is attracted.

For many years, it was assumed that the forked tongue of reptiles functioned as an organ of touch or hearing. Experimental work has demonstrated that these reptiles are endowed with a sense similar to that of smell, in which the tongue plays a major part. In the roof of the mouth are two groups of highly pigmented cells, called Jacobson's organs, from their first describer. To these, the forked tongue carries for identification odorous particles picked up from the air or the surface of objects. In the several series of lizards observed, it was shown that the tongue-Jacobson's organ mechanism was not only of value to the forked-tongue lizards in the finding of food, but that in egg-brooding species it assisted materially in identifying the home nest and in keeping track of the eggs.

The American skinks are among those that brood the eggs. The nest is ordinarily underneath bark or wood in fairly damp, well-rotted debris. Instead of leaving the eggs after laying, as do a great many reptiles, the female skink stays with them throughout the incubation period. There are short intervals away from the nest for hunting and perhaps a sun-bath now and then, but usually she is on hand to ward off any stray marauder of not too great a size from an approach to her treasure.

The common skink of the eastern and central states, *Eumeces fasciatus,* is known as the "scorpion" in the southern parts of its range and is popularly believed to be venomous. In color and form, *Eumeces fasciatus* is subject to a great deal of variation. New-hatched *fasciatus* are almost black with five yellow lines running lengthwise along the body; the tail is a brilliant, enamel blue. In the warm southern states, *fasciatus* loses its lines and dark color with growth and gradually becomes a light reddish-tan. The head of males, especially, becomes swollen about the temples, markedly separate from the neck, and takes on a bright shading of orange. A large mature skink so slightly resembles the young form that the two were formerly assumed to belong to different species. In the northern parts of its range—New York and Massachusetts—it does not reach the size of southern specimens and retains through life a slightly dulled version of the infant pattern and coloring.

A number of other species of Eumeces extend the range of the genus throughout all except the more northerly parts of the country. All are similar in general appearance, character and habits in nature and in captivity.

The skinks are largely residents of localities where a ground covering of rotting wood and vegetation retains moisture. In the cage, imitate these conditions in two-thirds of the container with well-rotted leaf mold or peat moss, kept slightly damp and layered over with bark and moss. The remainder of the cage should be floored with clean sand for a feeding and sun-bathing area.

Meal worms and wax worms provide such insect fare as the skinks need, but in addition, a beaten egg and chopped meat mixture should be fed twice weekly.

THE ALLIGATOR LIZARDS

In several species, similar in habit and appearance, the lizards of the genus Gerrhonotus range through the Pacific Coast states, the Southwest and southern Texas. Somewhat like the skinks in general form, short and weak-legged with long bodies and obtusely pointed heads, the Gerrhonotus are usually called alligator lizards. The name is not only a reference to form and movement, but refers also to the large and heavily keeled scales arranged circularly about body and tail of most species. A conspicuous mark of the group is the fold of soft skin running along each side from the head to the hind leg, bordered above by the keeled, square scales of the back and by large, smooth belly scutes below.

The alligator lizards, like the skinks, move partly by means of wriggling movements of the body and do not attempt to use the short, weak legs for support. Although not to be classed for speed with sprinters like the collared and zebra-tailed lizards, they are by no means slow and their slithering and evasive motion is an effective means of avoiding capture. In attempts at escape, the alligator lizard often loses its tail and the twitching discard, diverting the pursuer, makes the attempt successful. According to species, the alligator lizards live in a variety of terrain—the deserts, the coastal regions of the West, the interior valleys and the lower mountains. Although reptiles of sandy and rocky regions, they are sometimes found in gardens and cultivated areas.

The alligator lizards have a forked, rather heavy tongue, like that of the skinks, which functions in the same manner. When hunting food, the tongue is repeatedly thrust out and touches objects encountered.

In nature, their food consists of almost anything in the way of insect life they can catch with their limited speed. Beetles and their larvae are a standby. In addition, they are not averse to a meal of small lizard if it be available and, consequently, should not be caged with smaller species.

Remarkably hardy, they need a dry cage with plenty of hiding places of bark or rock, a moderate amount of sun and a mixed diet of such insects as are available.

THE "GLASS SNAKE"

Closely related to the alligator lizards, the genus Ophisaurus, with its three North American species, is of particular interest. Commonly called the "glass snake" or the "joint snake," Ophisaurus represents the extreme degree of the tendency to substitute wriggling movements for the use of legs. Ancestral forms seemingly found legs useless and perhaps even inconvenient in their mode of life with the result that in their descendants, they have been entirely discarded. This loss of limbs, accompanied by the development of a tail about twice the length of the body, has given Ophisaurus a snakelike form and appearance. The uninformed always regard it as a snake, although obvious characteristics such as movable eyelids and a rigid, undivided lower jaw definitely place it as a lizard.

Ophisaurus is very fragile and almost any encounter results in the long tail being broken, sometimes into a number of pieces. This protective device has given rise to stories of the fabulous joint snake with its ability to break into a number of fragments which reunite after the disaster. One of the earliest newspaper cartoons appeared in Benjamin Franklin's *Pennsylvania Gazette* just after the Revolution and concerned the great national problem whether to organize as one large nation or exist as separate small states. This famous cartoon was printed directly from a woodcut set with the type and represented a "joint snake" broken into thirteen fragments, each tagged with the initials of an original colony. The caption—JOIN OR DIE.

Like the skinks, the glass snake is hardy in captivity. In nature it is found under similar conditions, and the same type of cage and treatment will satisfactorily fill its needs.

THE RACE RUNNERS

The most colorful and active of the lizards of the United States are, fortunately, among those wide in distribution. Commonly called whip tails or race runners, the several species of the genus Cnemidophorus range from coast to coast throughout most of the central and southern parts of the country. A considerable variation in size and coloring is evident in the group, but the kinship of the several species is evident in the whiplash tails, long pointed heads, general coloring and behavior.

The race runners belong to that family of lizards known as the Teiidae, a New World family chiefly found in South and Central America and the Antilles. The Teiidae range in size from the small-bodied six-lined race runner of the eastern and southern states to the big and strong tegu of South America.

The six-lined race runner of the Mississippi Valley and the Gulf and eastern states, *Cnemidophorus sexlineatus sexlineatus*, is marked with a pattern of lengthwise lines along the sides and back. The ground color of the back is usually olive, brown or blackish, while the stripes, normally six in number, are yellowish and run from head to tail. The underparts of body and tail are clear, light blue or greenish blue, most intense on the tail and the underjaws. This is one of the smaller species, measuring about ten inches in total length, of which more than two-thirds is made up of the animal's long and slender tail.

The large race runners of the west coast and the Southwest, *Cnemidophorus tessellatus* and its several subspecies and related species, grow to as much as eighteen inches in length, of which about two-thirds is tail. Young specimens ordinarily display a striped pattern somewhat like that of the eastern species. As the animal matures, this lined pattern is obscured and broken up by the gradual appearance of yellowish spots in the dark areas between the stripes. These spots enlarge and fuse with the light areas, and the line effect is lost, giving the lizard a pattern of wavy light and dark cross bands. In one species of Texas and New Mexico, *perplexus*, the lined pattern is retained in maturity.

The race runners have no fondness for deep woods or swampy ground, but prefer semi-open, dry and sandy regions. They are gregarious in habit and, instead of being evenly distributed through a seemingly suitable region, tend to colonize.

Their favorite time for travel and hunting is in the morning and evening hours. Night and the heat of the midday find them snugly hidden beneath low-growing vegetation or resting either in their own burrows in the sand, or in abandoned tunnelings of mice, gophers or other burrowers. The race runners are often found in cultivated areas where the soil is of a light and sandy texture. In the San Joaquin Valley of California, colonies of them are found occasionally in orchards, fields and berry patches. Low-growing blackberry vines, ridged high, offer especially good cover for them and they are quick to take advantage of this prickly shelter. The race runner is fast and, by means of swift dashes and sudden changes of direction, almost defies capture.

Their natural food is composed mostly of all sorts of insects: grasshoppers, beetle larvae, crickets and the like. The tongue-Jacobson's organ combination is especially well developed in them and aids them in hunting concealed foods.

In captivity, in addition to flies, meal worms, wax worms, crickets, etc., feed twice weekly with an egg and meat mixture.

THE GECKOS

Because of nocturnal habit, shyness and delicacy, the geckos do not ordinarily find their way into the reptile collection. The name is the native term given to the large Asiatic lizards of this group in imitation of their nocturnal call. For, among the distinctive characteristics of the geckos, is a voice. With our own small species it is nothing more than a squeak. Other distinctive characteristics of the geckos are an elliptical eye pupil, soft finely granulated skin and in most species the possession of wide, adhesive discs on the toes for climbing smooth surfaces. Another unusual fact about them is that one can see through their heads. The arrangement of skull bones is such that there is no obstruction to light passing between the thin membranes covering the ear openings.

Of the four genera of geckos found in the United States, only two are native. *Hemidactylus turcicus,* a Mediterranean species, was introduced to Key West off the tip of Florida and has become established there. *Sphaerodactylus notatus,* a very small gecko commonly found in the Bahamas and Cuba, is also found in Key West and no doubt represents an emigration from the tropics.

Phyllodactylus tuberculosus, a species common in Mexico, ranges over the border into southern California. The last of the group, *Coleonyx variegatus,* the banded gecko, of the arid Southwest, is most worthy of consideration and, fortunately, the one most likely to come into the reptile fancier's hands. It was formerly considered very rare, but the advent of the automobile and consequent night traveling on the roads have revealed that Coleonyx is much more common than was suspected.

The banded gecko is marked by bands of rich reddish brown on a ground of white. The skin is so thin and fine in texture that blood shows through the unpigmented areas and imparts to them a slightly rosy hue. The impression this spectacular little creature gives is one of great delicacy and fragility. Fortunately, Coleonyx is hardy in captivity. Its quarters should be dry; floored with fine sand. Its habit is to come out to forage only at night and a good-sized piece of bark or wood must be provided for shelter during the day. Small meal worms, wax worms and baby crickets make up a satisfactory menu.

THE GILA MONSTER

With the exception of one species, our lizards are not venomous. Although many of them attempt to bite when freshly caught, few of them have the jaw-power or tooth-length to do more than pinch the skin. The exception to this rule is the Gila (pronounced Hee-la) monster, *Heloderma suspectum,* for years one of the old reliables of Sunday supplement natural history.

With its close ally of Mexico, *Heloderma horridum,* the Gila monster makes up the only family of lizards known to be venomous, although a similar form from Borneo is strongly suspected. The venom is secreted by glands situated in the rear of the bulging lower jaw and is transmitted to the bite through grooved teeth.

In color, the Gila monster is one of our most beautiful lizards. The bold patterns of large salmon-pink and black scales give it the appearance of being covered with heavy beadwork. In form, however, the Gila is ugly and clumsy. A sausage-shaped body, short stubby legs, thick rounded tail and blunt head combine to make a creature of no grace.

The Gila is unable to move with great agility and is inclined to sluggishness in captivity; but under the enlivening influence of sun and warmth, it has at times surprised persons inclined to take its previous quiet too much for granted, by the speed

Robert Snedigar

ALLIGATOR LIZARDS

William G. Hassler

"GLASS SNAKE"

C. M. Bogert

RACE RUNNER

BANDED GECKO

LEOPARD LIZARD

GILA MONSTER

of its short rush and snap. Although they permit handling after having become accustomed to captivity, these lizards should not be treated carelessly or casually. In the case of a captive Gila monster bite, the familiarity which breeds contempt breeds mostly contempt for the folly of the bitten individual. Only the very incautious and foolhardy provoke such incidents, and a captive Gila monster bite is certainly to be classed as an indication of incompetency in the handling of live reptiles.

The treatment of a Gila monster bite is essentially the same as that given for a rattler bite. Certainly the tourniquet and suction treatment is imperative. Shock and secondary bacterial infections are great dangers in a bite of this kind and must be prevented.

In captivity, the Gila monster should be kept in a *locked* cage of sturdy construction. Screen top and glass sides offer both ventilation and visibility. A rounded piece of bark large enough for the Gila to hide under, especially if the cage is at any time in the sun, is a needed furnishing. Large cacti, well-rooted and protected around the base with buried rocks, stand the punishment of this clumsy reptile's clawing and shoving. Smaller plant material is quickly rooted out and killed.

Of all lizards, the Gila is perhaps the easiest to feed. Eggs or a mixture of beaten eggs and meat are recognized as edible by the long forked tongue and eagerly lapped up. A Gila in good condition has a notably plump tail, round and heavy with stored fat which is called on in periods of food shortage.

For the southwestern lizards, some variation of the desert terrarium described on page 89 will be found satisfactory. For lizards from less arid regions, such as Anolis, a cage well planted with vines and shrubs that are not too delicate is needed. Sweet potatoes, sprouted and kept in jars of water or wet sand, give a quick trailing growth.

The most common ailment of captive lizards is a ricketlike disease which manifests itself in a number of fashions. In geckos, the jaw softens and the lizard is unable to bite vigorously. In other species, the eyes become inflamed and pus-filled. Still others display a tendency to slow movement and a partial paralysis of the hind legs. These symptoms are the beginning of emaciation and death unless the trouble is checked. Actual infections and organic disease may at times show these same symptoms, but the reptile keeper is fairly safe in assuming that his animal's difficulty is due to diet deficiency. Treatment is simple and, combined with good feeding and care, cleans up the condition if it has not progressed too far.

The lack of adequate sunshine is the great cause of this trouble. Well-planted, well ventilated cages receiving plenty of sun, will do much to make curative treatment unnecessary.

Sun is also a powerful killer of reptiles. In no case allow a poorly ventilated cage or a cage without adequate shade areas to remain in the sun for more than a few minutes.

Ultra-violet light treatments serve as a partial substitute for sun, if a lamp is available for a few minutes once or twice a week. Certain bulbs on the market which screw into an ordinary light socket give off a limited amount of ultra-violet and may be burned over the cage more or less continuously without harm.

Like our other animals, the lizards have their own parasite problems. Lizards from the South, particularly the fence lizard and Anolis, are often badly infested with a tiny red mite. In the Anolis, these pests cluster behind the insertion of the foreleg, around the vent and in any fold of skin. In the fence lizards, they sometimes infest an individual so badly that each scale seems to cover one. One remedy is to apply olive oil with a light camel's hair brush, working the oil well underneath the scales or into the clusters of mites in skin folds; one application is enough. If Dri-Die is available, it does a better job.

If his stock is properly housed, healthy and well fed, it is more than likely that, at the appropriate season, the lizard keeper will find nests of eggs in his cages. Although with species like the egg-brooding skinks, it may be just as well to leave the eggs with the mother, in general it is safest to remove them and take care of them. Sterilized peat moss, very lightly dampened, is the best medium for incubating eggs. After placing the eggs on clean, slightly moist sand in the bottom of a dish, cover over with the peat, and place the dish where a constant temperature of nearly 80°F. prevails. Watch carefully for signs of drying out or mold. If the latter occurs, change the dish and contents. Most lizard eggs hatch in about eight weeks.

This same method has proven satisfactory with turtle and snake eggs.

Snakes

AN ACQUAINTANCE with snakes speedily makes this maligned class of animals a favorite, if not *the* favorite group of the live animal fancier. With few exceptions the non-venomous snakes of the United States are beautiful in color and pattern, lively and interesting in behavior and readily tamed and kept. It is unfortunate that such a handsome and attractive type of life should be cursed with the heavy burden of dislike which superstition has cast upon reptiles and amphibians generally and the snakes in particular.

Snakes inspire every human emotion except indifference. People who dislike them hate them with an intensity they accord no other creature. People who like them often regard them (especially when it leads them into making lap and house pets of venomous species) with what may mildly be called an excessive affection. The common horror and dread of snakes is inspired by man's sorry experience with deadly serpents and the endless store of misinformation and legend built around them. There is room in the mind of man for a new conception of the relation of reptiles to himself. Because the lion and the tiger are dangerous jungle kings, their humble relative the house cat is not denied a seat by the fire. The dog's place in his master's affection is not jeopardized by the ravages of his cousin the wolf. Why then should a hundred harmless snakes be destroyed because the rattlesnake is deadly? As man discovered his friends and found out his enemies among the other animals, so he can, with a little trouble, learn to appreciate the value and beauty of the many harmless reptiles in spite of the menace of a harmful few.

There are about one hundred and thirty species of snake found within the borders of the United States. Of this number only four types are venomous. The others are no more dangerous or harmful to man than any small woods or field creature capable of inflicting a scratch or a harmless bite. Unlike the greater number of small animals—field mice, rabbits, ground squirrels and other small game—snakes are not our economic enemies. The greater number of them are our economic friends and of substantial, although not often appreciated, aid in keeping down both insect and small mammal pests.

There is probably a larger fund of misinformation concerning snakes than any other group of animals. In fact, most of the popular ideas about snakes are ingenious and violently defended bits of decidedly unnatural history.

These fables range from the patently absurd statement that the bite of a snake can kill a tree and has the potency to produce a quick and devastating dry rot in the seasoned wood of wagon wheels and plow tongues, to the less obviously absurd but equally untrue belief that female snakes swallow their young in times of stress to protect them.

Breath is wasted in trying to discredit the first. The second represents a natural history fable vouched for by eminent observers of early days and still popularly believed to be not only possible but probable snake behavior. The studies of the anatomists reveal no provision for this protective measure of the young. Physical evidence points to a certainty that any snake entering another snake's gullet has simply gone the way of all flesh. There *is* truth in the cases reported of a swarm of young issuing from the broken and battered body of a large female snake. The story is usually encountered in connection with the rattler, a snake which gives birth to living young. However, instead of having taken refuge in the mother's stomach, the young, MacDuff-like, have been from their "mother's womb untimely ripped," and their forced coming to light is premature birth and slaughter in one.

Another favorite legend of wide and established belief with meager foundation in fact concerns the ability of snakes to "charm," respectively, small mammals and birds for food and human beings for spite. The "human being" part of the story seems an unnecessary and unduly melodramatic note. The "charming" of birds is, observed *carelessly*, a fact. But, observed carefully, there is a slightly different story. Intruders in the vicinity of the nests of many birds, such as the killdeer, are met with a peculiar and characteristic behavior. By a considerable display of agony and distress, the bird attempts to simulate, rather skillfully, too, a crippled condition which suggests to human being, dog, cat or snake, that if pursued she will prove an easy prey. Human beings are fooled by this trick, and so are dogs and cats. The snake, on the other hand, seems to be able to resist the impulse which will lead him away from the vicinity of the nest. The bird, frantically over-acting, trying to concentrate the reptile's attention on herself, approaches too close and is caught.

Many organs have developed in the reptiles in a fashion quite unlike the evolution of the same structures in other animals. Descended remotely from four-legged animals, nearly all snakes have entirely dispensed with the limbs of their ancestors. Naturally, with the shrinking of the legs and their final disappearance, other structures had to take up the job of locomotion. In birds, mammals and even the snakes' close relatives, the lizards, the ribs are rigid and serve as a protective cage for the delicate internal organs.

In snakes the ribs, while still retaining their protective office, have in addition a part in the task of locomotion.

Each pair of ribs is matched with one of the wide belly scales or scutes. These serve to brace against the roughness of bark or ground and the snake is pushed ahead as muscular waves rippling along the ribs move the scutes in a walking fashion. Obviously, to progress in this fashion the body of the snake (with the exception of head and sometimes tail) is flat on the ground or limb at all times and is never disposed in the series of vertical loops depicted so often in conventional representations.

Another misrepresentation of snake posture is familiar to everyone. Although not a physically impossible feat, it is not the habit of arboreal snakes to twist in symmetrical coils about the limb of a tree like string around a stick. When climbing or resting, snakes seldom encircle a branch with anything more than a curl of the tail, but drape their length along the bark in a series of clinging loops. Disposed in this fashion, the snake can, by a quick twist of the body, drop from its perch to the safety of water or concealing and protective shrubbery.

Although snakes have no external ear opening, the organs of hearing within the head are complete and there is no doubt that, by means of bone conduction, they pick up sound to some extent. Their hearing is not as acute or discriminating as the pictures of flute-playing snake charmers would have us believe. Actually, it is most unlikely that these serpents hear anything of the charmer's quieting melodies.

The "charming" and the swaying movement of snakes in supposed time to music is the response to another sense entirely—sight. The snakes are interested in the circling and rhythmic motion of the charmer's flute.

For years I kept at home a snake of placid and quiet disposition, whose favored station was the top of the radio on my desk. He liked the warmth of it and showed no more elation or appreciation of Beethoven and Brahms than annoyance or excitement over football games and election returns.

Most snakes see things in terms of movement rather than form and color. Any slight motion in the vicinity is rewarded by their interest. It is this accent on sight in terms of movement that makes it possible for a frog to "freeze" and lose a pursuing garter snake. This often happens in a cage where several snakes are being fed at the same time and the cage is well permeated with the frog scent. The senses of taste and smell are of little value in finding the prey, and if the frog keeps absolutely still the snake may even touch him and not find him. But if he so much as blink or twitch! . . .

In snakes, as in other animals, sight is of many degrees of acuteness. The burrowing species have less need for sharp sight in their hunting underground or in the galleries of rotted logs than the snakes which hunt in the open during daylight hours for elusive and swift prey. The coachwhips and the related racers are examples of this latter group. Researches into the structure of their eyes and that of snakes of similar habit have shown that the lenses are tinged with yellow of the same value as that used in the modern sportsman's shooting spectacles. As the sight of the rifleman is sharpened and rendered less subject to fatigue by his glasses, so the eyes of these snakes are rendered more acute by the colored lenses.

Aside from the facial pits of the rattler and his kin (see page 179) and the similar functioning lip pits of certain of the boas, the most interesting and peculiar sense mechanism of the snakes is the tongue. Even today—not only in the minds of country folk but in those of people who should know better—the slender forked tongue of the snake is the "stinger," and its flickering motion is supposed to be inspired by a vicious desire to poison the very atmosphere. Poor snake! In reality, his tongue is of all tongues the least harmful.

Its function is to pick up information, particularly as to things edible and inedible. Diverse opinions as to the exact manner in which the snake used its tongue were current among naturalists for many years. Some authorities stated that the organ was intended as an imitation of a wiggling worm to decoy hungry birds within striking distance. Others believed it to be an organ of touch similar in function to the antennae of insects, while still others said it was an organ of hearing.

Actually, as in the forked tongue lizards, the tongue of the snake picks up odorous particles from the air and carries them back to the two sensory bodies in the roof of the mouth (Jacobson's organs) for identification.

Although primarily of greatest use in the trailing of food, the tongue of the snake is also used for the questioning of mysterious and unfamiliar quantities such as man, and is perhaps also an aid in the searching out of a mate during the breeding season. When at rest the tongue is retracted into a tubular sheath in the floor of the mouth. It is not necessary for the snake to open his mouth in order to protrude the tongue; a notch in the upper lip permits its free passage.

Just as birds molt and appear in new plumage, so the snakes shed at more or less regular intervals and reveal, each species in its own degree, the maximum of beauty in color and texture. A snake about to shed gives the first indication of its intent by retiring to some quiet corner and remaining there, sluggish, oblivious to food or distraction. The skin looks rough and lusterless and, in snakes customarily glittering with a prismatic sheen, all brightness has disappeared. The eyes become cast over by an almost opaque skim-milk whiteness and, in snakes marked on smooth belly scutes with black, the spottings become obscured by the same milkiness. Most snakes at this time show a marked desire to soak, and a good-sized water container should be available to them.

When the new skin beneath is ready, the eyes clear up overnight. The two layers—old and new—are no longer stuck together but have separated and the snake again can see. A couple of days—usually spent in soaking—elapse before the snake actually sheds. This final step in renovation is started by the snake coming out of its quietness and restlessly stirring about; rubbing the snout and lips against rocks and bark, loosening the edges of the old skin around the mouth. The rest is easy. By crawling through twigs that catch the old skin or slithering between rocks or even under and over the bodies of companions, the snake crawls right out of his old covering. The cast in a healthy snake is usually complete even to the clear shields over the eyes. These inside-out cast-offs are a common find in field and wood. Shedding is intimately associated with growth but is not entirely dependent upon it. Snakes shed whether growing or not.

The fact about snakes which has met with the least distortion in the folklore of the past and in the lurid accounts of the present-day journalists is that they are carnivorous. An exception to this is the fiction that snakes have a liking for milk and that the milk snake, *Lampropeltis doliata triangulum,* in particular, makes a practice of stealing from unsuspecting or complaisant cows. In spite of the fact that a good-sized milk snake can be coiled into a pint cup and still have room, some farmers credit this slight creature with the ability and willingness to consume a couple of gallons of milk in one session.

Thus, in the majority of cases, when a snake is found in the vicinity of barns, he is vindictively slaughtered and his bright carcass thrown to the hogs to guarantee the safety of the milk supply. If the campaign for milk protection be vigorously waged, another must soon be instituted. Rats and mice enjoy destructive revels in the granary, hay loft and store rooms and must be dealt with by cats, traps, or by poison. When harmless snakes are found about buildings and in cultivated fields, the farmer does well to pass the time of day courteously and give the guest a word of welcome. For, although uninvited, the snake is there because he is needed, and when his rodent game becomes scarce will move on to more fruitful hunting grounds. It is true that certain of the larger snakes are likely to take an occasional young chick or duck or to indulge in a meal of fresh eggs. Such lapses are small charges to make against their great return in vermin-catching. Rats kill and eat young chicks, steal eggs, ravage the winter's grain and vegetables and by their filth and waste destroy great quantities of food in addition to the amount they consume. The snake's quiet efficiency in reducing the numbers of these pests should be accepted gratefully and the occasional loss of an egg or chick cheerfully overlooked.

These hunters of small warm-blooded game kill their prey by constriction. Simultaneously with a lightning-quick and bulldog bite, the coils of the body are thrown around the animal in a tense grip. The pressure is sudden enough and strong enough to stop the breathing instantly. Probably also, the pressure upon the heart is sufficient to check its action and quick and painless death without struggle or great outcry is the result.

The tension of the coils is maintained for some moments, apparently until such time as the body of the reptile detects no heart beat or tremor in the prey. Then it is slowly released and, after a preliminary exploration, the snake engulfs the prey without mastication or tearing apart.

This ability of snakes to devour easily creatures many times the mass of their own heads is the result of a highly specialized arrangement of the bones of the skull. This skeletal structure is one of the great anatomical differences between the snakes and the rest of the animal world.

The lower jaw of the snake is divided into two entirely separate sections instead of being in one rigid piece as in other animals. The split is at the front and the two halves of the jaw, each of which is capable of independent action, are held together loosely at the chin by extremely flexible ligaments and skin. The upper jaw likewise is split at the front and its sections are capable of forward and backward movement, but to a much less degree than the bones of the lower jaw. In feeding, after constriction the snake usually noses about until he finds the head of the prey. Then, often after a preliminary gape to loosen up the muscles of his jaws, his mouth stretches unbelievably wide and clamps down upon the animal. Both upper and lower jaws and, in addition, the movable bones of the roof of the mouth, are well provided with sharp, backward-pointing, curved teeth and these hooks firmly engage the meal. Then one side of the head loosens its hold and is moved forward and the teeth re-engaged. A rhythmic contraction of the jaw muscles pulls the food a little farther into the mouth on that side. The other side of the head repeats the action in its turn and in this fashion the prey is "walked" into the reptile's throat. As swallowing goes on, the bones of the snake's head, especially those of the lower jaw, spread apart to an astounding degree. The gullet, too, is seen to be enormously elastic. If the prey is large, the outer skin of the snake stretches so as to separate widely the normally close scale rows.

If the meal is small, the swallowing is over quickly; if large, the action may be long-drawn-out. A laborious feeding often demonstrates to the careful observer another highly specialized mechanism peculiar to snakes. As the teeth of one side are loosened and moved forward for another pull, there is seen in the front of the mouth a slightly protruding tube lying between the body of the prey and the bottom of the mouth. The pressure of a large mouthful upon the normal breathing channels of the snake naturally inhibits, if not entirely checks, the passage of air to the lungs. This tubelike upper part of the windpipe and its extension into the forepart of the mouth is nature's answer to the problem.

Swallowing, once started, becomes almost an automatic process and will be continued as long as the jaws have something to work on. This necessitates a careful watch when feeding water snakes and black snakes in groups. If two individuals grasp the same tidbit, often not only the food, but the tenacious smaller snake as well, is devoured by the larger snake. Sometimes that is the end of it. Cannibalism is no lapse from virtue among them and such a fortuitous addition to a meal, if not too large, is welcome. If too large to handle, the eater disgorges the eatee, sometimes in a moribund condition but often very little worse for wear.

The digestion of snakes is remarkably efficient. The digestive fluids dissolve the entire carcass of the prey almost completely and the only recognizable remains to appear in the excretion are horny and acid-proof structures.

The efficiency of his digestion and the fact that his body runs no heating plant and hence needs material only for growth, reproduction and the energy used in movement, gives the reptile an advantage over the warm-blooded animals. During a given period his food needs are proportionately much less. By restricting movement to a minimum, a snake can draw upon comparatively small quantities of stored fat to carry him over periods of fasting which warm-blooded animals could not survive.

This ability to fast underlies one of the snake-keeper's most difficult problems. Hunger in most captive animals soon becomes strong enough to overshadow natural fear at accepting food from the hands of man. Not so with reptiles. While some species take to captive feeding readily, others refuse steadfastly to be interested in anything put into the cage and eventually starve to death in the presence of food without apparent display of hunger or obvious distress. If the fast is broken by force feeding and careful force feedings are continued at regular intervals, the snake will sometimes voluntarily begin to eat.

It is well at least to make the offer of natural food before resorting to the extreme measure of force feeding. In the matter of natural food, a decidedly ticklish question and one that has been provocative of much argument and complaint arises. Natural food, for snakes, implied living food, and without it many of them cannot be expected to survive long. The average person's dislike for snakes and the unthinking sentimentality of a part of the public have led to many expressions of indignation and disapproval of the sometimes necessary practice of permitting the snake to kill and eat his food in his own fashion.

If the plaintiff in these cases can, along with his or her expressions of disgust and self-righteous horror, produce a well-attested affidavit that he or she is a strict vegetarian and has no taste for the white meat of chicken or the juicy richness of prime ribs, he or she may be allowed a slight basis for argument and protest. Otherwise, the case is not valid and should be thrown out. If the plaintiff, as has happened, is a lady wearing a fur coat, spare no feelings.

Death for small animals and birds in the coils of a constricting snake is practically instantaneous and lacks the sickening and bloody brutality, the passing shriek of agony and dying moan which accompany man's own sport and food-getting.

Unfortunately, all snakes are not constrictors. Death in the jaws of the water and garter snakes and the black snake is not a pleasant sight, as they use their bodies only to hold the struggling prey while they devour it. However, we are too apt to read into the behavior of lower animals our own human emotions and physical reactions to circumstance and, hearing the shriek of a water-snake-caught frog, instantly think of the physical pain, the mental anguish and horrible realizations of a human being in like extremity. Cruel and painful as its situation must be, the frog's cry and struggle for escape are not those of a sensitive and intelligent creature but the instinctive reflexes of a creature incapable of feeling pain in the human sense of the word. The greater part of the frog's agony exists in the spectator's mind.

Being chicken-hearted, afraid of dentists and averse either to bearing or inflicting pain, I prefer to get around the need *if possible* rather than place reliance upon the above argument to drown dying squeaks and croaks and to still conscience. Snakes, particularly the non-constrictors, do not invariably demand living food. Many of them can be induced to take dead mice, frogs, birds or even pieces of lean beef or fish. If the snake is adjusted to cage life and feeding normally, the switch to killed game can sometimes be made by dangling the food on the end of a string in front of his nose. The movement and the food scent often stimulate him into taking the meal. A few such lessons lead the reptile to a point where it will immediately take anything that is offered without further teasing.

Force feeding, when necessary, should be carefully and cleanly done. For most snakes, lean beef cut in strips is satisfactory. A wooden forceps with polished and rounded points is best for larger snakes. A good pair can be whittled down from the wooden forceps sold in photographic supply stores for the handling of wet prints and film. These have the advantage over metal of not being so likely to break teeth or injure the delicate tissues of mouth and gullet. Dip a strip of meat in cod liver oil or even in milk for lubrication and, holding the snake behind the head, gently force the food into the mouth and down the gullet. Feed several small pieces instead of one large one. When the food is in the gullet, easy and slow pressure above it will move it down into the stomach. Of course, during such feeding the snake's body must be supported and no strain put upon the delicate spine. Feed on a table or on your lap.

For small snakes, the wooden forceps are likely to be too clumsy and heavy, and metal forceps will have to be used. These should have rounded points and must be used with caution. Earthworms instead of meat are good for hand-feeding small snakes as the forceps may be sheathed in the worm itself and the snake thus protected from internal scratches or punctures. An alternative method of feeding young or small snakes is to administer a semi-liquid mixture of beaten egg and chopped meat through an eye dropper tube with its opening enlarged. A wooden piston to press the food out is easily made.

Long-continued force feeding is almost certain to result in injury and death and should not be done unless really necessary. Liver and fish are more difficult to handle than meat but they should be alternated with it. The ideal force feeding for larger snakes is dead mice dipped in cod liver oil; or, a rat may be cut in pieces and fed to the snake in the same manner as meat. While the matter of reptilian dietetics has not received a great deal of attention, the natural habit of snakes in consuming only whole animals makes it reasonable to suppose that such a meal contains just the right balance of minerals, proteins and vitamins and is the exact formula for keeping them in good condition.

Under especially favorable conditions of captivity, snakes sometimes mate and bring forth young. In the wild, courtship and mating take place in the spring. With species which hibernate in numbers, the scene is ordinarily the vicinity of the den in the days following the awakening of the group from winter sleep.

Besides those which lay clutches of tough-skinned eggs, a number of our snakes have living young. This difference in reproduction, while significant in that it perhaps represents the beginnings of a process of nourishing young within the mother's body, after the manner of the mammals, is but one of degree. Although recent research has demonstrated that there is much more to the story, for the sake of simplicity we may consider that essentially all the snakes are egg-layers. In the "live-bearers," the eggs are retained within the mother's body during the entire incubation period, protected from chance enemies and unfavorable conditions of heat and humidity. So far as we know, the eggs receive no nourishment from her other than the oxygen necessary to life. This is diffused through the egg envelope to the embryo from the tiny blood vessels lining the oviduct. Hatching takes place immediately upon presentation to the outside world as the young break out of the membrane-like sac.

In egg-laying snakes, the clutch of eggs is deposited in the hollow of a stump, under sheets of bark or in some other sheltered location, and left to hatch. After egg-laying most reptiles display little or no interest in the fate of their offspring. As far as the meager records indicate, snakes are not exceptions. Some species, however, are known to coil about and protect their eggs and it is not unlikely that study and observation of reptile habits will show the trait to be present in more species than we realize.

At the time of laying, the eggs of snakes, unlike those of birds, have already undergone a considerable development within the oviduct of the mother, and are well advanced upon presentation to the outer world. Their exposure for some weeks to varied and ofttimes adverse conditions of humidity and heat has made it necessary for the embryos of the egg-layers to be provided with a heavier and tougher envelope than the thin sac of the live-bearing species.

During incubation, moisture is absorbed from the surrounding damp earth or rotten wood and the egg becomes considerably enlarged. The once regular outlines become distorted and lumpy and the outlines and even the coloring of the baby snake are easily seen through the parchment-like shell.

Young snakes are equipped with a special implement, the egg-tooth—a tiny, sharp projecting tip to the snout—for use in effecting an escape from the shell. This egg-tooth is very similar to that found on the tip of the bill of baby birds, and like it is shed within a short period after hatching. The manner of use is somewhat different. A bird or chick uses the egg-tooth as a breaking instrument and "pips" the shell by repeated blows. The egg-tooth of the reptiles is used as a cutting edge, and quick rakes of it on the inside of the shell cut an opening large enough for the infant to crawl out. Some hours elapse after the young snake slits the shell and pokes an inquisitive head out before it is able to emerge and desert the useless hull. Often, at this time, ill-advised aid is given the youngster; he is forcibly removed from the shell, and his reason for not being in a hurry becomes apparent. A portion of the nourishing yolk still remains attached to the navel and it is his intention, desire and need to remain quiet long enough to absorb it.

The delicate membrane encasing the young of live-bearers hardly needs the slash of an egg-tooth to liberate the infant. Yet, it is a matter of record that in many such species, this now useless appendage has been retained.

THE BULL SNAKES

The most economically important group of snakes within the borders of the United States is undoubtedly the gopher, or as they are called in some regions, the bull snakes.

Large powerful constrictors preying upon such small destructive rodents as mice, rats, squirrels, gophers and even rabbits, these snakes earn their keep in agricultural regions and are worthy of the grudging protection they are beginning to receive.

In irrigated country especially, gopher and squirrel burrows in ditch banks are the start of wash-outs costing much money and labor to repair as well as a heavy expense in damage to crops from the lack of water. Without realizing it, many a farm boy of these districts (the writer with a shamed face and a contrite heart includes himself among them) has paid heavily for the dubious pleasure of killing a gopher snake. Many hours of back-breaking, hurried labor to save precious water are wasted because of the activity of rodents—activity that the presence of a snake at the right time would surely have prevented. It is with a belated gratitude that I record the defiant residence in a corner of our alfalfa field of a large gopher snake which managed to elude all our attempts to slaughter him. Having been warned of the venomousness and danger of all snakes since infancy, the bulk and sheer magnificence of this creature intimidated us and we were not inclined to tackle him without adequate weapons. Sticks and stones were fortunately hard to

PRAIRIE BULL SNAKES EMERGING FROM EGGS

PRAIRIE BULL SNAKE

John C. Orth

PINE SNAKE

Robert Snedigar

BANDED WATER SNAKE

find and invariably by the time we had located a safe implement of death, he had slithered his length (we estimated it as at least twelve feet, but it probably was closer to six) into a convenient burrow or into the green safety of the alfalfa where we dared not follow.

Although bull snakes are really harmless, it is not difficult to understand the rancher's fear of them. The common species of the Middle West, *Pituophis catenifer sayi*, and the pine snakes of the east, *Pituophis melanoleucus* and *mugitis*, attain a length of more than eight feet and when disturbed and in danger display an alarming and formidable hostility. In addition to a menacing attitude, a vicious willingness to strike at the interloper, and a rapid vibration of the tail which in dry weeds or leaves produces an unpleasant rattler-like whir, the bull and pine snakes have the power to expel the breath violently in a hiss of hair-raising magnitude. In some localities this characteristic has given them the name of "blow" snakes and the breath is regarded as poisonous, foul-odored and blighting.

The largest of the gopher or bull snakes, *Pituophis catenifer sayi*, is found through the Middle West in the area lying between the Rockies and the Mississippi. Individuals of more than eight feet are not uncommon and one of this length will measure six or more inches around the body. The head is small compared to the body and has often been likened to that of a turtle. In profile it is seen to be less flat than that of most snakes and the nose is less blunt—almost pointed.

The coloring of the bull snake is subject to a great deal of regional variation. In most specimens a ground color of rich yellow, sometimes almost orange, is overlaid by a row of squarish blotches of dark reddish brown along the back and a series of smaller lighter markings along the sides. The skin is not smooth in appearance, but looks as if plaited out of thin straw, for each dorsal scale has a lengthwise ridge or keel. The underparts of the snake are yellowish, spotted at the sides with a mottling of black.

Reaching its maximum in numbers and size in Texas, the bull snake is succeeded in the West by the smaller variety, *Pituophis catenifer affinis*, the Arizona gopher snake, similar in character and appearance. On the west coast, the Pacific gopher snake, *Pituophis catenifer catenifer*, appears. Although local reports attribute to it a size equal to the bull snake of the prairies, and old-timers insist that in the days when the San Joaquin Valley was one vast grainfield, snakes of this species eight and nine feet long were common, a Pacific gopher snake six feet long would be decidedly out of the ordinary.

In most cases the bull and gopher snake adapt themselves readily to cage life. An Arizona gopher for some years was a resident in the American Museum of Natural History. After living for some months contentedly under what turned out to be the inappropriate name of "Oscar," this snake refused food and shortly thereafter showed unmistakable signs of physical distress. An examination revealed that "Oscar" was a female and that she was unable to rid herself of the unwelcome and useless burden of a clutch of unfertilized eggs. An anesthetic was given and, by the surgeon's art, "Oscar" was successfully delivered. Two years after the event, with a six-inch scar to show, she lived and participated silently but with interest in all discussions of her operation. Incidentally, the name became Esmeralda.

THE PINE SNAKE

The pine snake, *Pituophis melanoleucus,* takes the place of the bull and gopher snakes east of the Mississippi and along the Atlantic seaboard. In form, size and general character, pine snakes strongly resemble their western relatives. They are similar in pattern, too, but quite different in color. Pine snakes from the barrens of southern New Jersey, the northern part of the range, have a ground color of grayish white on the back; on the sides and belly it becomes a pure polished white. The dorsal blotches are black or nearly so, running together and spattered with the whitish ground color in the head and forepart of the reptile. In the southerly forms of the pine snake, the black is replaced with a reddish brown and the blotches are more diffused. In them the white is duller and less clear.

In all this group of the bull and pine snakes, there seems to be a protective correspondence between the coloring of the reptile and the prevailing tones of the environment. The northern pine snake with its white and black is lost against the white sands and dark vegetable debris of the barrens. In the dry pine woods of the South, the reddish tones of the southern form merge with the needle-covered forest floor. The blotched yellow of the bull and the gopher snakes, melting into the confused patterns of shadows, grass clumps and dun earth of prairie or sand of arid country, serves admirably for concealment.

Generalizations on the behavior of captive animals are notoriously unsafe; individuals of species commonly supposed to be vicious and intractable often become docile and tractable. Likewise, ordinarily good-natured species may refuse to be gentled and like nothing better than a nip at the hand that feeds them. Snakes perhaps present this equation of the individual in a lesser degree than most animals, but no certain predictions can be made as to the quality in captivity of any species. In the matter of gentling and feeding, the reptile keeper often finds his guests behaving in a manner totally at variance with the recorded experience of others.

The large bull snake is each year caught in enormous numbers for sale to circuses and sideshows in which, especially in the East, it appears under all sorts of exotic and venomous aspects. Under these difficult conditions, many of the snakes are exceedingly gentle, feed readily upon rats, mice or eggs and manage somehow to survive. Imagine the disappointment of a snake fancier who, a couple of years ago, bought a large bull snake. The snake refused to eat or be comforted and noisily resented handling. Eventually, because of its fasting, force feeding was necessary. As almost inevitably happens when force feeding is carried on over long periods of time, the creature developed a case of mouth rot and died, still protesting at being kept in captivity.

Pine snakes have often been described as sullen and resentful, indifferent feeders and prone to suicide by starvation. The statement undoubtedly has ample foundation in the experience of the persons expressing the opinion, yet the experience of others has been that pine snakes are tractable and readily adjust to cage life and feeding.

In keeping bull and pine snakes, particular care should be taken to provide them with a dry hiding place where they can feel themselves secure. Without a small box, or a rockwork cave in which they may coil completely with only the head showing inquisitively at the entrance, they are unhappy and range up and down, pushing at the cage top, worrying the corners, trying to find a way out to some quiet and semi-dark seclusion.

For their food, rats and mice are in almost all cases acceptable. Most pine and bull snakes with a head large enough to take hen's eggs readily do so, and need no more than one or two eggs weekly. Smaller specimens undoubtedly could be well kept on a diet of bantam chicken or pigeon eggs.

THE GARTER SNAKE

The garter snakes are even more widely distributed through the United States than the bull and pine snakes. If no other snake finds its way into the small zoo, one or a dozen of these 'is certain to put in an appearance sooner or later.

In many parts of their range, garter snakes greatly outnumber snakes of other species. In some localities, such as the high mountain meadows of the Sierras, their number is almost unbelievable. Semi-aquatic in habit, the small snakes of this region frequent the borders of the shallow meadow ponds and, as the passer-by disturbs them, take to the water in such hordes that a small pond fairly boils with their agitated movements.

It is to be regretted that with their wide distribution and great numbers, the garter snakes should be of little economic value. They are powerless to do actual harm to man and certainly under some circumstances consume a share of destructive insects. But against this meager credit is a heavy debit for their inroads on the numbers of frogs and young toads, powerful natural agents in insect control, and a bill for the destruction of young game fish.

The kinship of the garter snakes and the water snakes is attested by the habit of both of voiding an offensive, acrid-smelling secretion when frightened. In captivity, after this preliminary unpleasantness and a little snappishness, they become gentle and cease to strike or otherwise offend.

The basic characteristic of most species of garter snakes is the three light stripes running the length of the body against a dark ground color. This may or may not be accompanied by spots along the sides; sometimes the side stripes are faint (in one species they are entirely absent). The ground color between the stripes may be solid or broken into bars or checkered patterns. The variations on the basic theme are seemingly infinite and have been for years the joy and despair of the classifiers.

Of our many species of garter snakes, the common garter snake, *Thamnophis sirtalis,* and its numerous subspecies have a nationwide distribution and a correspondingly wide range of color and pattern variation. In the East the ground color is ordinarily black, brown or olive and the stripes greenish, dusty yellow or yellow. Between the central stripe and those of the sides, a row of rather symmetrical squarish dark markings is present, although in the dark specimens it may not be obvious. The body of the snake is moderately stout and the head distinct from the neck. Young specimens are more slender and vivid in color. The scales of the back are lightly ridged. The common garter snake very rarely attains as much as three feet in length.

Among the more beautiful color phases of this species is a form from Florida with belly, chin and under-parts of the head bright and clean blue-green, black ground color of body and slightly green straw-colored stripings.

In several of the varieties of the common garter snake, the central stripe is stronger or different in color from those of the sides, a row of rather symmetrical squarish dark markings is one subspecies, *Thamnophis sirtalis infernalis* of the Pacific coast, the side stripes are entirely absent in adults and from this characteristic the snake takes its usual name of one-striped garter snake. *Infernalis* presumably relates to its erratic behavior and the impression it makes on man. Ditmars notes this behavior as being the distinctive character which sets it apart, in his mind, from related species. When disturbed, *infernalis* strikes viciously and, in an attempt to escape, throws its body into a series of sidewise loops, at the same time keeping a menacing head in the direction of the enemy. If the snake is trying to get up a steep and sandy ditch bank, the impression that it is about to attack is complete. Only one member of our family, an old tom cat by the unheroic name of Henry, did not fear these supposedly dangerous reptiles. He killed and ate them. We ascribed his eventual death in a fit to this morbid habit and diet.

The usual western garter snake belongs to another group, that of *Thamnophis ordinoides* and its numerous subspecies. It is like *sirtalis* in general form, color, pattern and the infinite number of variations upon them, but is smaller in size, and has generally duller colors.

Among the more distinct and less difficult of identification is Butler's garter snake of the Middle West, *Thamnophis butleri*. Stout of body, with a narrow head and neck, *butleri* is smaller than *sirtalis*, seldom running more than eighteen or twenty inches in length. In habit, *butleri* is very gentle, easy to feed, and although not particularly bright in color or beautiful in pattern, is an eminently suitable member of a mixed terrarium family. With toads large enough not to be tempting, a wood or box turtle and perhaps even a frog or two, several *butleri* help to make an attractive terrarium picture.

The care of garter snakes is simple. If frogs and small toads are available in sufficient quantities, the larger snakes will be better off for the balanced ration the whole food affords. While small toads are acceptable, larger specimens are not favored and often will be disgorged after having been swallowed. Presumably the greater development of the skin glands and the quantities of poison exuded are the unpleasant factors. Garter snakes of the less aquatic species relish earthworms and these, in addition to being the staple food for small individuals, offer a means of conditioning larger ones to an easily prepared and always available diet. The training is simple. Mix a few earthworms in with a small quantity of hamburger or other chopped meat and place in the cage in a shallow dish or on a piece of glass. The snakes, attracted by the wiggling and scent of the worms, attack the mass. Often one or two such meals are enough and thereafter the snakes are willing to dispense with the stimulus of the worm movement and scent and will take meat alone. Beef should not be fed entirely but should be alternated with chopped raw liver and fish. After the snakes have become well-conditioned to such diet, cod liver oil or the contents of a broken vitamin capsule may be added. Feed once weekly.

Cages for garter snakes should be airy and light. Plenty of clean dry hiding space under pieces of sheet moss and bark should be provided. As these snakes have a marked fondness for soaking, a shallow container of clean, fresh water must be provided. If possible, place the cage where sunlight will reach it for a part of each day, especially during the winter months.

THE RIBBON SNAKES

Closely allied to the garter snakes is the ribbon snake, *Thamnophis sauritus,* and its southern and western subspecies. Like them, it bears the characteristic three stripes along the back and sides. In body, the ribbon snake is the slimmest of the garter snake group and one of the most delicately formed. The tail is long, about a third of the body length. The ribbon snake is not a large snake either in bulk or length; mature specimens usually average about two feet.

Throughout the considerable range covered by the ribbon snake, it displays a great deal of variation, as might be expected. In the subspecies of the Southeast, *Thamnophis sauritus sack-*

eni, the back stripe is either entirely lacking or very faintly marked in the forepart of the body only.

More aquatic than the garter snakes, ribbon snakes frequent the vicinity of streams and ponds and feed upon small fish, salamanders and their larvae, frogs, small toads and tadpoles. They seem seldom to venture afield and appear to have no interest in insect life or earthworms. As a result, the problem of conditioning them to a diet other than their natural foods is slightly different than in the case of the garter snakes.

The solution, however, is essentially the same. Raw fish, cut into small strips or pieces and dropped into the water container of the cage, will sometimes be picked up and eaten without ado. If stimulation is needed, the addition of a few small live fish or tadpoles will usually suffice: in pursuit of living prey, the substitute food is encountered and, with the idea perhaps that they have caught the fish they have been chasing, they devour it greedily and cheerfully. In later feedings, the live stimulus may be dispensed with.

In this connection, it may be noted that snakes condition to this type of feeding better when kept in large groups than when kept singly or in small numbers. By their example the intelligent, daring or, perhaps, merely hungry individuals, stir up interest in the food among their less active cage mates.

Cages for ribbon snakes need plenty of clean water and dry hiding places. Welcome additions are clumps of growing grass or other vegetation; and a dried shrub or well-twigged limb to provide a roomy sunning place.

THE WATER SNAKES

The water snakes of the genus Natrix are probably the least attractive in habit, color and behavior of all our non-venomous snakes. As their popular name indicates, they are commonly found in the vicinity of water and prey upon the small cold-blooded life of ponds and streams. Their destruction of frogs, toads and fish makes them emphatically not of economic value.

Their semi-aquatic habit limits their distribution to regions where streams, marshes and lakes are abundant and therefore the group is not well represented west of the Mississippi Valley.

In body, water snakes are stout and heavy, with coarsely keeled and rough-looking scales. In adult specimens, the color pattern is indistinct and dirty-looking except in newly shed individuals. At that time, some species display markings and colorings of great beauty and delicacy.

Representative of the entire group in habit and behavior is the banded water snake or "moccasin," *Natrix sipedon*. Its distribution is the most entensive of all and with its several subspecies, it covers the area lying between southern Canada and Florida and the Atlantic and the Rockies.

The only portion of this species to retain a brilliant and fine coloring in maturity is the underpart. There, red and black dots and irregular blotches are strikingly accented by a clear white or yellow ground color. The beauty of the color is increased and intensified by the polished smoothness of the belly scutes just as the patterns of the back are obscured by the coarse, heavy keeling of the scales.

These snakes are difficult to catch as they seldom stray far from the refuge of water and plop into it at the slightest hint of danger. Once in its safety, they easily evade clumsy attempts to catch them. They are usually encountered draped in the bushes above the water, basking in the sun. If cornered and unable to reach safety their behavior is such as to warrant the fear and dislike in which the group is generally held. The head flattens into a perfect semblance of the triangle commonly held to be the mark of a venomous snake. Short, vicious strikes in the direction of the foe give warning that here is a business-like scrapper. The body loses its meager roundness and flattens close to the ground in a series of S-loops; the air becomes filled with the acrid reek of the fluid voided from glands within the vent. The bite, an imbedding of the large, strong teeth and a quick, succeeding backward jerk, tears the flesh, and sometimes results in severe lacerations, particularly on a finger or thumb. Such a bite does not show the two large punctures inflicted by the genuine water moccasin, *Ancistrodon piscivorus*, with which the water snakes are continually confused. (See page 194.)

Because of this chronic confusion, any study collection of reptiles should include such of the water snakes as are available, in spite of their unpleasant qualities, if only to define the differences between these non-venomous species and the dangerous true moccasin.

Most species of water snake adjust themselves to feeding in captivity and take readily to a diet of raw fish with an occasional meal of frog. One feeding a week is plenty. Confined to small cages, singly, with only a small water container, they can

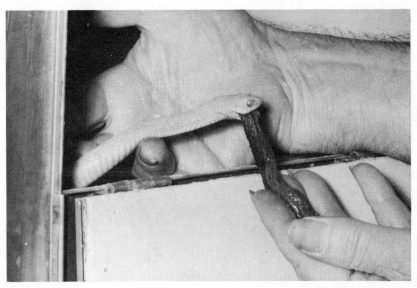

Robert Snedigar

RARE ALBINO GARTER SNAKE BEING HAND-FED WITH MEAT

Robert Snedigar

FOUR-LINED RAT SNAKE

Robert Snedigar

FOX SNAKE DINING

Robert Snedigar

CORN SNAKE

be kept for long periods but do much better in large airy cages or pits in company with their own or related species. A good-sized pool for soaking and feeding is not essential but is appreciated. In addition to limbs and brush to climb into for sunning, it is necessary that they be assured of protected dry areas on the ground of the cage. Without sunshine and with a perpetually damp cage bottom, water snakes are apt to succumb to a pneumonia-like disease.

THE RAT SNAKES

The snakes of the genus Elaphe, known generally as the rat snakes, are represented in the United States by ten species and a number of subspecies, most of which are large and handsome snakes in maturity.

In the field these snakes are of prime importance in the destruction of vermin. Powerful constrictors and agile hunters, they loose the wrath of an insatiate hunger upon the small rodents infesting plantations and farms. In corn and grain fields, the vicinity of buildings, anywhere that their small game is likely to nest and breed, there the Elaphe are found. In the fields and woods ground-nesting fowl and their eggs may suffer from the raids of these snakes, but not enough to warrant their extermination for the sake of the birds. As has been mentioned earlier in this chapter, rats are relentless killers of small poultry and wreck enormous havoc in the nests of such birds as pheasants and quail. The snakes efficiently help control the rats and lessen their ravages to an extent that leaves a generous credit to their account.

The Elaphe do not bear living young but deposit a clutch of from ten to twenty-four eggs, depending on species and size of the laying female, in moist and protected locations such as the inside of rotten logs, stumps, or under large pieces of old bark. The eggs are not left exposed but are covered with rotten wood or leaf mold and hatch in from six to eight weeks.

It is in the newly-hatched young that the kinship of the several species is most evident. Young pilot blacks, *Elaphe obsoleta obsoleta,* instead of being solid and somber in hue like the adults, are light in color and pattern and strongly resemble adults of the related variety, the gray rat snake, *Elaphe obsoleta spiloides,* a spotted snake found in parts of the South. In the same manner, the young of the four-lined rat snake, *Elaphe obsoleta quadrivittata,* instead of showing the four lengthwise stripes that give the species its name, emerge from the egg with a gray color and dark spots. In both these color- and pattern-changing species, about two years elapse before adult coloring is attained. During that time the young rat snake presents an interesting exposition of the manner in which the differently colored and marked species evolved from a common ancestry.

THE PILOT BLACK SNAKE

The pilot black snake, so called because of a superstition that it acts in some vague and ill-defined capacity as "pilot" and guard to the venomous copperhead and rattler sharing its range, frequents shelving rock ledges and sunny slopes. It is often encountered along old stone walls lining country roads and in planted fields and orchards—especially abandoned orchards left to the birds and the insidious inroads of mice and moles.

When mature, the pilot black is one of our most imposing snakes. Pattern it has none, unless the faint remains of its infantile spottings on the thin skin between the scales can be called pattern. Its color is a rich and lustrous black. The scales are small and slightly ridged but not sufficiently so to destroy the effect of a smooth and polished skin. A distinguishing characteristic of the pilot black and the Elaphe generally is the shape of the body. Instead of being cylindrical and rounded as in many snakes, a cross section would be somewhat the shape of a horseshoe. The almost angular head is covered with large smooth shields which abruptly, and with a sharp demarcation, give place to the small body scales. While the pilot black is one of our largest snakes, measuring as much as eight feet, it hardly attains the size with which it is often credited.

In some states of the South, the pilot black is called black rat snake in recognition of a principal prey.

THE GRAY RAT SNAKE

The gray rat snake, *Elaphe obsoleta spiloides,* a subspecies of the pilot black, is like it in form, scalation and habit, but retains in maturity the silvery gray and brown of infancy. The gray snake is less apt to resent capture than most, although it is capable of putting up a vicious battle if inclined to be nasty. Often it may be picked up by the roadside as casually as a piece of rope, but this is no absolute guarantee of future behavior. All the Elaphe are a bit unstable temperamentally and the gray rat snake is no exception. One day I had occasion to enter a large and well-planted cage in which a group of pilot blacks, yellow and gray rat snakes, water snakes and large garter snakes had made their quarters for some months. Except at feeding times, this cage was seldom disturbed. When formerly confined in small cages, all these snakes had been quiet and docile and with the exception of two yellow rat snakes, had never shown fight. My visit was one of inspection. Accustomed to house-cat behavior from them, I was startled, as I reached above me to take a large pilot black from a limb, to have a smaller pilot black and a gray rat snake make simultaneous strikes for my face from their resting place on a shelf high up in the back of the cage. Their action was an automatic response to the surprise of seeing my head and shoulders rise to the level of their secluded perch. Abrupt and jerky movements tend to produce such reactions in snakes and should be avoided. In this case the roominess of the cage, the sun flooding the roof of the greenhouse and the security from disturbance they had so long enjoyed had perhaps to some extent re-converted them to wildness.

My own response was entirely automatic. I fell off the box on which I was standing into the big water pan in the bottom of the cage, much to the consternation of several easily upset water snakes soaking there.

I went away and came back a half hour later, re-entered the cage somewhat noisily and mounted the box in less apparitional fashion. Without the least offensive move or attempt to escape, this time they permitted me to handle them freely, proving—at least to me—that the original fight reaction was the result of surprise and that all blame for the incident was really mine.

THE FOUR-LINED RAT SNAKE

The four-lined rat snake, *Elaphe obsoleta quadrivittata,* has a less extensive range than the pilot black, but is a common and abhorred familiar of the farm, especially the poultry yards, in its chosen territory of the southeastern and Gulf states. Much like the pilot black in body and scaling, this handsome snake bears on a ground color of greenish yellow, the four dark lines from which it takes its name.

THE CORN SNAKE

The most strikingly colored and marked of all the Elaphe is the corn snake, *Elaphe guttata*, a common and fairly abundant snake of the southeastern states. Individuals vary, but in a typical specimen, crimson saddle-markings on the back, each outlined with a scale's width of black, are brilliant against a body color of light red. The sides are blotched with two series of less regular and lighter markings. In some cases the edges of the upper row tend to blur and merge into a lengthwise band similar to that of the yellow rat snake. Ordinarily the lower sides are much lighter in color than the back and make a decided contrast to the mosaic-like pattern of sharp black and pure white on the belly scutes.

The corn snake is quiet and slow to anger and often may be caught with little fuss and excitement. Around barns and yards, the species seems to adapt itself readily to the comings and goings of people and if not hounded, tends to lose its natural shyness and timidity. Camping behind an old barn in South Carolina while on a visit to the swamps bordering the Savannah River, we saw an excellent and very gentle specimen crawl out of a breast-high crevice in the logs just above our noisy breakfast table. The weather had been cold and wet and the early morning sun was the best treat we had had for days. The corn snake seemed to agree and, oblivious to our noise and movement, lay on the narrow ledge gratefully soaking in the warmth. He seemed not to mind us—even, perhaps, to like us. As he had shown what we chose to construe as an interest in joining our party, we graciously acceded—and grabbed a bag—and join it he did.

THE FOX SNAKE

More than any other of the Elaphe, the fox snake, *Elaphe vulpinus,* uses the protective device so common among reptiles, of emitting an ill-smelling fluid from special glands at the base of the tail. The very name of this snake, fox snake, refers to this heavy "foxy" scent. Fortunately, after a few handlings, it refrains from using this device and ordinarily becomes a favorite in the reptile collection.

The fox snake lacks the spectacular beauty of the corn snake. The prevailing tone is a somewhat somber grayish brown or yellowish brown, regularly marked with rich brown spots along the back and alternating smaller blotches along the sides. The belly is yellowish, slightly mottled with a darker value. The head is inclined to reddish and is marked by stripes or blotches. One of the most obvious characteristics of the fox snakes is the abruptly pointed and heavy tail. The body is heavier and stouter than in the other Elaphe and is perhaps a reason for a less arboreal habit.

THE KING SNAKES

The most widely known and admired characteristic of the king snakes is a trait that in other animals is regarded with distaste and abhorrence. The king snake is a cannibal with an appetite that reduces even the venomous snakes it encounters to just so much more to eat. The very name, king snake, is a popular recognition of this group's ability to battle and kill the rattler and his kind successfully and of its apparent immunity to the effects of snake venom. Indeed, the venomous snakes, dependent upon poison for protection, and slow in movement, perhaps offer a less difficult conquest for their traditional enemy than some of the more active harmless snakes.

The king snake's body, strong, well-proportioned and almost cylindrical, is admirably suited to dealing death to other snakes by constriction, but this cannibalistic fare is a side issue in his diet. From the many examinations of stomach contents that have been recorded and the experience of every keeper of live king snakes, it is known that the principal and favored food of the group is small warm-blooded game such as rats, mice and gophers, and that, therefore, the king snake is of great economic value.

With exceptions, the king snakes are quiet in temper and lack the jittery disposition of some of the other harmless snakes of our country. Larger kings, especially, refrain from violent attempts to escape or vicious resistance to capture. However, if hurt and angered, they sell their lives as dearly as any cornered creature. Fortunately, the prowess of the group in killing venomous species has made them exempt from man's hatred in most localities.

This prowess, like every fact concerning snakes, has been subject to great distortion and exaggeration. A caucus of rolling stones composed of ex-cowhands, miners, a tramp printer, two truck-drivers and an accordionist, jumbled together in the same California bunkhouse, devoted each evening to an open-minded discussion of anything that came to hand. The bars were down and any story, disguised by bluster and sufficiently supported by vigorous pounding on the table, could masquerade as fact and be welcomed.

The discovery one night by the agitated Chinese cook that a king snake about eighteen inches long had made its home in the messhouse woodpile, brought out loud and extensive expositions of "facts" concerning these snakes. Ordinarily, in this little gathering, discussions followed a fairly definite routine. A statement was met with a counterstatement—followed by loud argument. The next step was innuendo with sly allusions to personal qualities and deficiencies. The debate usually wound up with the direct insult and invitations to and from everybody to step outside and settle the matter.

But, on the subject of snakes, these rolling stones abandoned all precedent and every tale (each taller than the last) was supported by a hearty "Amen" chorus. This unexpected accord and unnatural agreement were disturbing to me, but my gentle hints at doubt and disbelief were, like the small voice, abruptly stilled.

"Smart kid, huh? Think *you* know so much. Sure, he could eat a six-foot rattler. Sure, he could. A little old king snake can eat the biggest rattler a-goin' . . . and with no trouble. Where's he put 'im—in his stomach—they does it! Quick as a piece of rattler gets in his stomach, it's eat up with the juices —dissolved. Stick your finger in a king snake's belly and you wouldn't have no finger. That's how fast it works. That's how he eats the big 'un!"

According to this same symposium, the king snake kills its reptile prey variously be biting them, by stretching (an ingenious method in which the king snake wraps an upper coil about the rattler's neck, his tail around the rattler's middle and than by an abrupt and rigid straightening of the slack in his own body rips the victim in two) and by hypnosis. The accordionist was particularly taken by the latter theory. He begin a complicated history of the evil eye and the efficacy of the amulet of blue and yellow glass he dragged from under four undershirts. Loudly shushed, he subsided into a series of hurt protests, accompanied by plaintive minor harmonies from his accordion, that it was not in the least his intention to go around equipping rattlesnakes with blue and yellow charms against the king snake's malignant glare.

Of the several species of king snake found within the United States, one species in particular, *Lampropeltis getulus,* is represented in several subspecies from coast to coast. The wide variation in color pattern—a pattern fundamentally one of light bands (white, creamy-white, yellow or greenish yellow) upon a dark ground color (dark olive-brown, blue-black or greenish brown)—found within the species over its considerable range and diversity of habitat, offers a fine example of the adaptations of an evolving and migrating species to the needs of new environments.

The several subspecies of *getulus* are much alike in body proportion. The head is slightly distinct from the neck, small, rounded and with little angularity. The body itself is cylindrical, tapering slightly from the middle toward head and tail. The tail is short and terminates in a horny tip. The undersurface of wide scutes if flattened and meets the sides in a rounded angle. The scales are unkeeled and highly polished. Newly shed specimens, especially in sunlight, display a prismatic sheen.

The first to come under the eye of the white man, the eastern representative of this group, *Lampropeltis getulus getulus,* bears a greater burden of popular names than its western and Mississippi Valley relatives. Its first describer, Catesby, in 1731 christened it the chain snake, "from some resemblance to a chain that seems in many cases to environ the body." Other names such as cowsucker, black moccasin and master snake, are obvious in their origins. Such names as horse-racer, thunder snake and thunder-and-lightning snake are more obscure and doubtless point to supposed qualities and forgotten incidents in the history of our contact with these snakes.

Similar in basic coloring and character to the chain king snake, but much different in the effect produced, is *Lampropeltis getulus holbrooki* of the lower Mississippi Valley. Its popular names, guinea snake and speckled king snake, although less used than the plain and indefinite term king snake, well express the color scheme. Instead of solid colors in more or less regular markings, the speckled king snake, like the guinea fowl with its light and dark spotted feathers, sports scales of rich blue-black spotted with white or yellow.

Along both sides and back, at regular intervals, the spottings of the scales are larger and tend to run from one scale to another. This produces the effect of the definite chain design of the eastern *getulus*, especially when viewed from a little distance or with half-closed eyes. The belly of the speckled king snake is blotched black and yellow.

The speckled king snake is fairly common within its range,

very similar to the chain king in habit and foods and, like it, adaptable to captive conditions.

It is quite gentle but not overburdened with intelligence. Apparently it recognizes nothing except its appetite, which is impartially satisfied in nature with snakes of many species, lizards, mice, rats, birds and the eggs of reptiles and birds. In captivity, it prefers small rodents and the less strongly scented snakes, and, after a little teaching, relishes pieces of raw beef.

Serious doubt and a skeptic eye have been cast upon the jungle traveler's tales of man-eating boas and pythons. Such stories are today rejected with scorn and vigor by the knowing. It is therefore with considerable reticence that I venture to state that snakes *will*, with a singular lack of appreciation of the difference in size, occasionally attempt to devour a full-grown human being. If picked up and handled by a person who has within moments been handling other snakes, rats or mice, his majesty the king snake is more than likely, upon getting the familiar food scent from the holder's hand, to jump simultaneously at conclusions and a finger. If permitted to proceed, the regular chewing motions "walk" the finger into his mouth and unmistakably demonstrate his intention and his lack of judgment. This lack of size discrimination may partly explain the generally accepted pugnacity of the king snakes and the occasional cases in which they attack and kill snakes too large for them to swallow.

The black and white king snake of the West, *Lampropeltis getulus californiae,* displays two dissimilar patterns within its range. One is a pattern of creamy-white rings separated by double their width or more of dark chocolate brown or black. In the other, the coloring is the same but instead of a ringed pattern the snake is marked by a light stripe, usually broken in a few places, and running from head to tail, on the brown ground color. The two were for many years classified as separate species, *Lampropeltis getulus boylii,* the ringed phase and *Lampropeltis californiae,* the striped. Broods of young hatched in the San Diego Zoo in which both striped and ringed occurred as well as the puzzling in-betweens formerly assumed to be hybrids of the two "species," proved the two to be identical.

THE MILK SNAKES

Smaller in size than the *getulus* group, and differing from it in that red enters into the pattern, is that group of king snakes

BLACK KING SNAKE

Robert Snedigar

FLORIDA KING SNAKE

CALIFORNIA KING SNAKE, STRIPED PATTERN

CALIFORNIA KING SNAKE, BANDED PATTERN

typified by the common eastern milk snake, *Lampropeltis doliata triangulum.* Cylindrical in body, flat-bellied and with head but slightly distinct from the neck, the milk snake is more slender than the other king snakes, and its pattern one of spots rather than bands. Adults are commonly between two and three feet in length, and are marked by a series of saddles of brown or reddish brown, extending far down the side, on a dark speckled ground of gray or reddish gray.

The saddles are encircled with a narrow width of black. On the sides are smaller roundish spots, also black-edged, alternating with the saddles, while a still lov. .r series of almost entirely black spots laps over the right angle of the sides and belly. The belly is white, checkered with black in a pattern somewhat reminiscent of the corn snake's tessellated scutes.

Through the central and eastern states, this snake is a common find. Most often it frequents upland regions away from swamps and streams and, a tireless hunter of the rats and mice which form its principal prey, is often found in the vicinity of buildings and in the farmyard. Hayfields are a favored haunt and it is often found, belly distended with young mice, quietly resting beneath a shock of hay. True to the king snake tradition, the milk snake includes smaller snakes and lizards in its diet. The young and eggs of birds sometimes fall to its share, but its habits are not such as to make these any considerable item, and the great number of pests it destroys should make it a welcome guest in any farmer's yards and barns.

In captivity, the milk snake and its subspecies are generally less satisfactory than the larger king snakes. Although occasional specimens will, after a period of adjustment, take to feeding, most of them steadily refuse food and eventually perish of suicidal starvation.

The eggs of the milk snake average from eight to ten, and are laid in midsummer in rotten wood, under bark, in leaf mold or even in manure piles. The shells are white and have somewhat the smooth and satin look of kid leather. The young hatch in early fall and might just as well be turned loose in a favorable spot soon after hatching, as they are most difficult to raise.

In the Middle West, the typical milk snake intergrades with another subspecies, the red milk snake, *Lampropeltis t. sypsila,* a form widely scattered through the upper Mississippi Valley. Smaller than the eastern milk snake, this member of the *triangulum* group has wider saddles of dark red against a lighter ground color. The saddles are edged with black as in the eastern form, but when seen from above the snake has the appearance of a ringed snake rather than that of a spotted one. Thus it represents a transition stage between the blotched milk snake and the ringed king snakes.

THE RINGED KING SNAKES

Belonging to several species, the ringed king snakes have in common similar patterns of red, black and yellow or whitish yellow. Because of its similarity to the markings of the venomous coral snakes, this pattern often leads to the ringed kings being confused with these dangerous reptiles. There have been cases in which amateur collectors, believing they had captured and were handling one of the non-venomous snakes, actually had picked up a coral snake.

"Mimicry" of this sort seems to be an established fact with other forms of life, especially in the insect world, but it is difficult to see what advantage to either could come from an imitation of the coral snake by the king, or the reverse.

These king snakes differ from the corals in the sequence of the bands. In the corals the red and black bands are separated by narrow yellow rings; in the king snakes the yellow and red bands are separated by black rings.

The scarlet king snake, *Lampropeltis doliata doliata,* is the smallest of the king snakes and seldom attains a length of more than eighteen inches. Because of its burrowing habit, it is less likely to be encountered than many rarer snakes of its range. A usual haunt is beneath the bark of fallen and decayed trees, where the blue-tailed skink and the fence lizards are likely to be hiding. In such places, there are also nests of young mice, and small snakes find refuge in them from the hot sun of midday and the chill of night. This king snake, although most handsome, adjusts to captivity indifferently well and its passion for concealment beneath the accessories of its cage makes it a poor exhibition specimen. In behavior it is quite gentle and not in the least vicious. When it can be persuaded to eat, it displays a preference for young mice. If a nest of these are quietly left in the cage just before dark and no later disturbance occurs, conditions approximating the circumstances of a natural feeding have been closely imitated and there is a good chance the snake will forget his unwillingness to eat.

The Arizona king snake, *Lampropeltis pyromelana,* grows

to twice the length of the red king snake, with a similar brilliancy and beauty of coloring. Instead of yellow, the light bands in this snake are whitish and the red is of a rich terracotta or brick red rather than scarlet. A distinctive character is the white nose.

Although from the Southwest, the Arizona king snake is not a desert dweller, as might be supposed, but prefers the mountainous regions. Specimens have been taken as high as 7,000 feet.

The many-banded king snake of California, *Lampropeltis zonata*, also prefers mountains to desert and is found at even higher elevations. This singularly beautiful snake is small, seldom attaining more than two feet, but is strikingly marked. The pattern is made up of many regular rings of whitish or yellow, separated from wide interspaces of brilliant red by black bands. As might be guessed from the brightness of color, this snake is a shy and retiring creature. Its distribution is largely restricted to the moister hills and mountain valleys of California. It has no great liking for dry regions and although found in both the Sierra Nevada and on the damp west slope of the Coast Range, is not recorded from the arid San Joaquin-Sacramento Basin lying between these mountains. Like others of the genus, its food is made up of small rodents, snakes and lizards. Old miners of the Mother Lode country credit it with astounding feats of rattler-killing and, although believing it also to be highly venomous, usually spare its life as the lesser of two evils.

THE RACERS

A considerable group of well-distributed species, specimens of which are almost certain to find a place in the small zoo, is that of the racers of the genus Coluber. Certainly in the eastern states, the black snake is such a frequently met (but less often caught) reptile that no collection with pretensions to making a display of the local small fauna could be considered complete without at least one.

Among the popular names listed for the king snake are thunder snake and thunder-and-lightning snake. It is to be regretted that these names, both of which fit the racers, should have been given inappropriately to another group. For the racers are greased-lightning. As for thunder, an unwise collector without gloves who manages to catch up with one can be depended upon to furnish thunder. Moderately well supplied with teeth, the racers know how to use them and encounters with them in the wild have an almost inevitable penalty of scratches.

The best and seemingly only practical method of catching these speeders is to make a flying tackle in the general direction in which the snake was last seen and hope that a wild grab will get him or that he will be pinned to the ground. Ordinarily, in catching a snake with the hands, it is advisable to take hold as close to the head as possible. In this case, however, there is no picking holds; the collector takes what he can get. If he gets anything, he gets plenty—often a great deal more than he bargains for. The nose is no suitable site for a snake bite and it is difficult to explain satisfactorily a lengthwise set of deep scratches along its prominence. Yet, in racer hunting, such lightly honored scars have been the result of unfortunate errors such as clamping down upon what seemed to be the head end of a small, partly concealed black snake and having it turn out to be the tail end of a big one!

Ounce for ounce, an irritated racer that wants to go elsewhere and can't has as much fury packed into his lithe and twisting length as any wildcat. The only thing the collector can do is to hang on and hope that the scars won't show. Fortunately, the teeth are comparatively small and the wounds, unlike the deep, ripped tears the larger water snakes inflict, are not much more than scratches.

The extraordinary fight and viciousness which the black snake, *Coluber constrictor constrictor,* shows when cornered has gained for this species the reputation of being irascible and pugnacious. This violent display is but the courage of desperation and indicates a real dependence upon speed and bluff for protection.

In spite of the scientific specific name of *constrictor,* and the general impression even on the part of naturalists, the black snake does not constrict its prey, but, like the water and garter snakes, kills it by eating it. The only way in which the black uses its coils in the killing or manipulation of food is to press the prey tightly to the ground while swallowing begins.

In captivity, black snakes are at first markedly nervous and have a tendency to worry back and forth seeking escape. Warmth seems to increase this activity and, on hot days, unless adequate hiding material is provided, torn and bleeding snouts from rubbing often spoil otherwise perfect specimens. Kept in large groups, they remain wild longer than if separated. Kept in singles or pairs and handled, first with gloves and then with bare hands after danger of nipping seems past, many specimens tame quickly and so completely that they will accept dead mice, frogs and even pieces of meat from the keeper's hand instead of the accustomed living foods.

The black snake shows a little more intelligence than the rest of its kind and when (and if) adjusted to a caged life, often surprises and gratifies its owner with unexpected displays of cleverness. It seems to have more individuality than most snakes, and even in a large group, certain ones quickly become distinct and recognizable because of what we might call personal qualities.

The black racer is among the very common snakes of the eastern states and ranges west into the Mississippi Valley. There it intergrades with the subspecies, *Coluber constrictor flaviventris,* which is known variously as the blue racer, the green racer and the yellow-bellied racer. As these names indicate, the coloring of this form is variable, but is generally much lighter than that of its somber eastern relative. The belly is always light, usually yellow but sometimes light blue-gray or greenish gray.

Like the black snake, the blue racer is active; a skillful climber, it is often to be seen high in the branches of shrubs and even trees.

The racers are carried to the west coast by *Coluber constrictor mormon.* which goes popularly under the same string of names as its Middle Western relative. This blue racer is smaller and more gentle than the black snake or the Mississippi Valley blue racer and in some localities shows surprisingly delicate and clear colorings. It is able to travel through the tops of shrubs and over brushy areas with considerable speed. For food it hunts small lizards, insects and snakes. Grasshoppers in season form a major item of diet, and the blue racer haunts wet meadows, grassy plains and prairies in search of them.

THE COACHWHIPS

Closely related to the above are the racers known as the coachwhips. These make up a separate genus, Masticophis. Two principal species of coachwhip, with many subspecies and a number of related forms, range from the Atlantic to the Pacific through the southern part of the United States. In no part of the range are they to be classed as common, although they are very well-known and the subject of much rural superstition. The slender build and great length of tail gives the eastern form, *flagellum,* very much the look of an animated rawhide whip. This appearance, coupled with the speed of the reptile, and its lashing viciousness when molested, have given rise to the country belief that the coachwhip pursues, catches and then "whips to death."

In habits and food, the eastern coachwhip is similar to the black racer sharing its territory, except that it disdains frogs and prefers to feed upon small rodents and birds. In temper it is markedly nervous and not inclined to be gentle, but there are exceptions.

The snakes of the *flagellum* group are fairly uniform in color, usually very dark on the head and shading to very light on the tail. The western *taeniatus* group inclines to a pattern of stripes.

THE BLACK GOPHER SNAKE

Any discussion of snakes with reference to their adaptability to captive conditions would be sadly incomplete without mention of the large black gopher snake of the southern states, *Drymarchon corais couperi.* This is one of our longest snakes and sometimes attains a length of well over seven feet. The body is fairly heavy. The scales are large, conspicuously smooth and polished, with the result that in sunlight newly shed specimens have a slight iridescence. In the East this snake is a glossy black or blue-black (one of its popular names is indigo snake) marked with mahogany red on the chin and throat. In Texas the color changes. There specimens begin to take on a little of the brown and olive tints of the tropical form of which the black gopher is the most northerly subspecies.

Although sometimes inclined to fight when first encountered and seemingly formidable because of its great size, the gopher snake is the species most to be relied upon for good nature and a ready acceptance of captive conditions. It has no food prejudices but willingly takes almost anything that is offered: frogs, small warm-blooded game, fish, lizards, smaller snakes and toads. Placid and with no hysterical tendencies, this snake freely submits to handling, and because of its extreme beauty and good nature a specimen proves a major attraction in a reptile collection.

THE HOG-NOSED SNAKE

Bluff is a large element in the defensive behavior of our harmless snakes. Most of them, if pressed, are willing to back their bluff with such teeth as they possess. Notable and almost laughable exceptions are the several very similar snakes of the genus Heterodon. In addition to the name of hog-nosed, given them from the peculiarly turned-up snout, this group has by sheer, unadulterated bluff gained a whole series of popular names of evil implication. Among them are spreading adder, puff adder, spread-head and blowing viper. The use of these is almost invariably tied up with a belief that the reptile is highly venomous.

Small, heavy and stocky of body and correspondingly slow of movement, the hog-nose is virtually defenseless. When discovered, escape is almost out of the question and the only resource left the snake is to frighten off the aggressor. To become successfully frightening, it makes use of several interesting devices and adaptations. The ribs high up on its neck are elongated very much as they are in the cobra and, by spreading them sideways, a hog-nose can flatten out a menacing and formidable hood. Its breath is expelled in violent hisses loud enough to be worthy of a snake twenty times its meager bulk. Short, stabbing strikes threaten the object of wrath. The whole attitude of the reptile is one of uncompromising hostility and virulence, well calculated to inspire terror and flight. I once knew a two-hundred-pound colored woman to get through the tiny window of a henhouse because a spread-head had appeared on the path outside while she was collecting eggs. The snake "came for her" as she opened the door. The eggs were a total loss; the henhouse was no better off for the experience, but— the snake escaped safely.

Given this vicious aspect, there seems to be no question as to what would happen if a person were unwise enough to put a hand within reach of the strike. Actually, if this *is* done, the strikes seem miraculously to fall short of the mark. A little teasing quickly demonstrates that the hog-nose cannot be persuaded to bite.

If the violent displays of pretended virulence do not succeed in winning freedom from annoyance, the hog-nose has another defensive trick. Touched by stick, stone or hand, he "dies" in extreme and terrible agony. Twisting and writhing, pretending injury, he rolls over and over, mouth wide open and snapping at his own body, till at last, with a final shudder, he is on his back, forked tongue protruded and still. Quite convincingly, the tail "dies" last in a pathetic and convulsive twitch.

The only defect in the performance is that it is too good. Turned over on his stomach, the hog-nose promptly flops onto his back again, apparently considering this to be the more realistic death pose.

It is to be regretted that after a very few displays in captivity, the hog-nose loses his taste for drama and refuses to act any more. Because of their good disposition, utter harmlessness, beautiful coloring and adaptability, snakes of this group are valuable additions to the live reptile collection.

The coloring varies with the species. The hog-nose of the East and Mississippi Valley, *Heterodon platyrhinos*, is ordinarily some shade of reddish or yellowish brown with dark irregular blotches or spots along the back. A totally black phase is found in certain localities. The western hog-nose, *Heterodon nasicus*, is paler in color than its eastern relative and has a more conspicuously turned-up snout. In habits it is similar.

The food of the hog-nose snake is the only black mark against the group. They live principally upon the beneficent and gentle toad. The skin poisons of amphibians, while offensive to most other snakes and poisonous to some, do not affect these reptiles and they are able to eat even such strong and unpalatable creatures as newts and pickerel frogs.

THE GREEN SNAKES

Certain small snakes are admirably suited to life in the terrarium. First among these are the green snakes of the genus Opheodrys. Two species are found throughout most of the United States lying east of the Rockies. The keeled grass snake, *Opheodrys aestivus*, sometimes called the green whip snake, is only to be distinguished from its relative *Opheodrys vernalis* by slightly ridged, instead of smooth, scales.

Both species are slender, delicate and inoffensive. Aside from speed and protective coloring, they have no defense whatever against death or capture. Their food is principally such large insects as grasshoppers and crickets.

COACHWHIP

BLACK RACER

C. M. Bogert

HOG-NOSED SNAKE

John C. Orth

GREEN SNAKE

THE RING-NECKED SNAKES

The ring-necked snakes of the genus Diadophis occur in many species throughout the United States. All are shy and secretive and are ordinarily found hiding beneath stones, old boards or other woodland shelter. A favored haunt is underneath the bark of dead trees. Consistent investigation of dead timber in regions in which ring-necks abound is certain to yield a reward. Small, slender and inoffensive snakes, the ring-necks are easily identified by the conspicuous circling of yellow just back of the head. The body color is usually some shade of dark gray, while the belly is in most cases yellow or yellow-orange.

Earthworms form the staple food of the ring-necks, but they are not averse to making a meal of smaller snakes or young lizards.

DE KAY'S SNAKE

The little brown snake, *Storeria dekayi*, is so small and obviously harmless that only the most prejudiced of humans go into a dither at the sight of it. Boys carry them in their pockets to school and are disappointed in the lack of resultant excitement.

Through semi-burrowing and hiding habits, the De Kay snake has been able to resist extermination even within very crowded and built-up districts and is occasionally found in vacant lots and unimproved areas of New York City.

Very small, seldom more than a foot in length, grayish brown with a double line of indistinct dots down the back, the De Kay is not difficult to identify. Its related species, *Storeria occipito-maculata*, with much the same range, has a conspicuously red ventral surface which gives it the name of red-bellied snake.

The two species are to be found throughout most of the United States east of the Rockies. Their distribution in this large area is not continuous. One or both may be abundant in a given locality; nearby they may be very difficult to find.

The food of Storeria consists almost entirely of earthworms and soft beetle larvae.

Snakes particularly dislike cages in which there is no protective shelter. Very excitable species, open to annoyance and light from all directions, worry themselves into a frenzy. A very little rubbing on a wire screen is sufficient to tear the end of the snout, and even on glass, persistent nosing will break the skin. Proper housing usually eliminates this problem, but new snakes should be watched for such traits. Covering over the cage completely sometimes helps. If the nose is broken or seems certain to become so, a couple of layers of adhesive tape may be placed over the end of the snout to save wear and tear. Be careful in applying tape not to plug the nostrils or the mouth. To get the adhesive to stick tightly, warm the gummed side with a match flame before applying.

The ideal snake cage has a plate glass front, a solid back and a lift up screen top with lock. The ends are of wood with a good-sized panel of stout wire mesh set in the upper half of each.

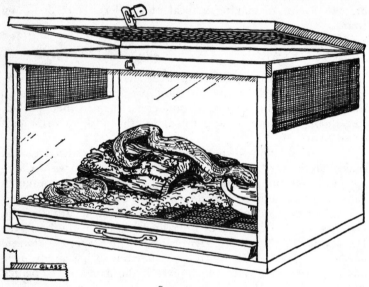

SNAKE CAGE

If the cage is large, the top may be in two sections. The glass front slides in channels so that it can be removed when the cage is to receive a thorough cleaning. The bottom may be of solid wood or of heavy and strong galvanized wire cloth laid over a crisscross of supporting wood or galvanized iron. The latter arrangement, while affording an adequate base for a layer of gravel, permits slopped-over water to pass through to the floor or a metal pan. Damp gravel is the great source of skin disease and pulmonary trouble in reptiles and this is one way of avoiding it. Concrete floors for cages are an abomination. They are abrasive, damp, cold and not fit for any reptile.

Snake cages should be roomy, light and well ventilated. Branches for climbing and sunning and a good-sized water container are essential. Many non-aquatic snakes welcome an opportunity to soak for a few days before shedding. If the cage is reached by sun, so much the better, if the caretaker keeps in mind the fact that reptiles have no way of ridding themselves of heat. Their temperature is always very close to that of their surroundings and direct sun on a hot day soon kills a snake. Death in this case is due to changes in the tissues produced by heat rather than, as was at first thought, to ultraviolet rays. This holds true also for lizards and turtles.

Cage sanitation is simple. Gravel and sand will have to be changed at intervals and replaced by new, but excreta can be scooped out with a spatula as deposited. Cage parasites are the most difficult problem. Small mites often occur in zoos in such numbers that they kill valuable specimens. The silica serogel dust insecticide, SG67, marketed as Dri-Die, already discussed in connection with bird parasites, is also effective in eliminating mites from reptiles. It might be well to put any new snake or lizard into a bag and give a good shaking up with the powder. The same treatment would be best for specimens known to be infested with parasites. Infested cages can be best cleaned with soap and water and painting the crevices and cracks with turpentine. Or the Dri-Die may be dusted in corners and on the cage gravel. This, however, can result in a film of the material being smeared on the glass of a cage front. If your animals are not on display—so what?

Much work is yet to be done on the internal parasites and diseases of reptiles. Out of many ailments we are able to diagnose and treat only a few of the more obvious ones. One of the most notorious of these is the infectious bacterial disease known as "mouth rot," from the characteristic ulceration of the inside of the mouth. This disease ordinarily starts in a snake which has injured itself by repeated striking at the cage front or which has been damaged by force feeding. Once started, the disease is easily carried to healthy snakes in other cages. Sick snakes should be isolated and extreme measures of cleanliness observed to check possible spread of the trouble. If the patient is a snake of a common species and of no particular value, the best policy, for the welfare of the collection, is to deal it a merciful death blow.

If the snake is to be kept, the first step is to cleanse the mouth. The disease attacks the tissues around the teeth and often fills the mouth with a foul-smelling, cheesy mass. This must be removed and the ulcerated surface beneath disinfected. Irrigation with a tepid solution of zephrin chloride is the best method. For this purpose, an ear syringe is a good implement. Continued washing will float out most of the pus, but it may be necessary to pick lightly around the teeth with a toothpick and cotton to get all. When well cleansed, wipe out the mouth gently with dry cotton and apply zephrin to the sore areas with a swab. Hold the snake's mouth open for a moment or two so the strength of the application will not be immediately reduced by mouth fluids.

The irrigation and cleansing treatment should be repeated daily until either the infection clears up or the patient succumbs.

Mouth rot patients must be handled with caution. The pus from the ulcerations contains many organisms and an untreated bite or scratch, even though the snake is not venomous, might result in a serious infection. Keep iodine handy and use it liberally.

Another common snake malady is a disease, presumably caused by a fungus, which produces blisters underneath the skin. If not checked these turn into bad sores and eventually become so large and so numerous that they kill the snake. Any bump or irregularity in the scaling of a new snake should be investigated. Sometimes a broken rib has healed awkwardly and forms a projection. These can be identified by feel and, of course, are let alone. After sterilizing the skin with iodine, an unidentified lump should be lightly slit with fine surgical scissors. If it is a fungus blister, there will be some oozing of fluid. Treatment consists of opening the slit the full length of the lump and swabbing the exposed pocket with full strength zephrin. The treatment of a badly blistered snake must be staggered over several days.

In many cases, especially when there are but one or two lumps, the operation reveals the presence within of a long white nematode worm. Treatment consists merely in the removal of the worm and disinfection of the area with iodine.

Snakes are for the most part gentle creatures and require much less taming than the usual wild animal. In fact, after having once been bagged, many specimens give up all hostile ideas and, at first perhaps a little nervously, permit handling.

Handling a snake is not the difficult matter the layman believes it to be. If the snake is new, caution and common sense advise that the neck just back of the head is the place to take hold. Lift gently from the cage, supporting the body with the other hand. Handle briefly and return. If the next time the cage is opened there is no sign of nervousness or escape, the head hold may be omitted. Fear of falling is the snake's greatest bugbear, and when held, the body must be well supported; otherwise twisting and threshing about ensue. Restraint and pressure are the things most likely to result in a bite when a snake is being held. Allowed to move freely from hand to hand, he believes that he is having his own way and is satisfied. Never permit teasing of snakes to induce them to strike.

The keeping of live animals daily presents new problems and situations for which there can be no rule-of-thumb guide. The keeper's best resource and the one on which he can most rely is his own good common sense. His use of it will be reflected in the welfare of his charges.

Venomous Snakes

ALTHOUGH VENOMOUS snakes are by no means advisable as pets, rattlesnake fever is a condition to which all reptile fanciers are susceptible. The disease occurs in varying degrees. Some individuals suffer a light attack and after going through the nuisance of caring for a rattler or copperhead for a few weeks, recover with no ill effects. One of the factors leading to quick recovery is a realization that friends and acquaintances are as much impressed by a couple of chicken snakes as they are by dangerous reptiles. Perhaps a little more. The harmless species put on a better show.

In cases of the second degree of the ailment, the patient, affected in a serious fashion, becomes attached to the group and always has one or two rattlers about. Still, others—these are the really bad cases—attempt to gentle and make pets of them.

The fact that venomous snakes can be gentled and brought to a state of tameness which permits of their being freely handled has received considerable publicity during the last several years. Stories of rattlers that perch on the owner's lap and assist in the solving of crossword puzzles make nice newspaper and magazine reading. Unfortunately the popular prints seldom run the parallel stories of these same pets and owners. Gruesomely detailed and grim hospital case histories, the latter come to print only in technical journals, if at all.

The need of seeming personally brave must be a fearful burden. When the continued exhibition of that bravery almost certainly involves painful and disgusting illness, permanent mutilation, perhaps death itself—a chicken heart is no liability. If you must keep venomous reptiles, don't be afraid to be afraid. Refuse to take unnecessary chances. If you feel that your exhibitionistic tendencies outweight entirely such fear as you can muster, give up the idea of snakes from the naturalist's point of view; get yourself a canvas pit with a carnival company and really bask in the shocked and startled adulation of an open-mouthed public.

The late Dr. Raymond Ditmars once stated in an interview that he had never been bitten by a venomous snake and that he would be very much ashamed of himself if he had been. The reason that he, like the greater number of leading herpetologists, had never had his health and pride so injured is that he had a proper respect for himself and a proper respect for his animal material.

Venomous snakes are not lap pets and no amount of sentimentalizing will make them so. Nor are they creatures to be airily dismissed. They are beautifully and curiously fashioned beings of varied and interesting habit, intensely entertaining to study and watch and—dangerous.

The greater number of our North American venomous serpents—our rattlesnakes, copperheads and the moccasins—are close relatives and belong to that group of snakes known as the "pit vipers." In captivity, some species do indifferently well while others, more adaptable, adjust themselves readily to the restrictions of cage life.

The distinguishing characteristic of these snakes which sets them off unmistakably from the harmless snakes of the same regions and which gives the group its name is the presence of a facial pit between the eye and the nostril. Deep and not opening into the mouth or the nasal passage, the pits are rich in nerve endings. It was obvious even to very early investigators that they were some sort of sense organ. However, for many years, the actual function of the pit remained as obscure as it was when the rattlesnake was first described.

Stated roughly and in terms of our own senses, the pits serve to pick up very delicate gradations of heat. From this information the snake is able to judge, as efficiently perhaps as we judge with the eye, the distance and direction of the warm object and is able, although deprived of all other aiding senses, to place a strike accurately when the object comes within reach.

The rattlers, copperheads and moccasins all have triangular, heavy-jowled heads, flattish bodies and strongly ridged scales. These characteristics are associated with them, but certain harmless snakes—the water snake and the hog-nose particularly—imitate them so well that they are of doubtful identification value.

Rattlesnakes are easily recognized by the jointed, horny appendage to the tail—the rattle—from which they take their name. In newly born rattlers, the tail terminates in a tiny button. Between this button and the tail of the adult snake, successive sheddings interpolate the rings or rattles. These are hollow, more or less symmetrical in shape and socketed loosely one into another. The interlocking is free enough to allow considerable motion. Vibration of the tail, by a rapid clicking of the segments against each other, produces the whirring noise.

It is rarely that perfect sets of rattles of any considerable size are found, because of their comparative fragility and the likelihood of their becoming caught between rocks or otherwise damaged. It would be unfair, however, to charge the snake with carelessness regarding his noise-maker. A crawling rattler does not drag the rattles on the ground but elevates the tail just enough to keep them up in the air.

The function of the rattle has always been problematical. Older writers surmised that it was intended to warn possible victims of the approach of a deadly enemy, but Darwin, with some apparent heat, dismissed this belled-cat theory:

> "I would almost as soon believe that the cat curls the end of its tail when preparing to spring in order to warn the doomed mouse. It is a much more probable view that the rattlesnake uses its rattle, the cobra expands its frill, and the puff-adder swells while hissing so loudly and harshly, in order to alarm the many birds and beasts which are known to attack even the most venomous species. Snakes act upon the same principle which makes a hen ruffle her feathers and expand her wings when a dog approaches her chickens."

The "dinner bell" theory of the function of the rattle, although never countenanced by scientists, has always been more or less popular. It has two versions. First: the rattle is used by the snake to imitate the stridulent love song of the grasshopper and the cicada and so betray these and other noisy insects into its voracious maw. It is unfortunate that rattlers, except perhaps when very young, do not feed habitually upon insects. Otherwise this is a pleasant and plausible idea.

The second "dinner bell" theory is that the rattle serves as a means of communication, a cheery tocsin call to all the tribe within hearing to gather for a light lunch. In rattlesnake country a proper willingness to be impressed and to listen quietly will almost surely bring out at least one old hearty ready to swear by the hot ashes of the supper fire that he, in person, has been the object of such a dinner call and only escaped from the encircling horde of hungry reptiles by a miracle. There is no point in mentioning at such a time that rattlers might gather by thousands and still not be able to dispose of any carcass too large for the greatest in the lot to swallow whole.

Actually, the motto "Don't Tread On Me" on the early American flags seems best to have expressed nature's intention when the rattlers were so peculiarly endowed. These ground-dwelling snakes have shared the same range of prairie and forest with heavy- and sharp-hooved animals since the continent was young. Any of these animals, startled by the warning whir, heed it quickly and well, and the serpent is no doubt often saved the futile use of its venom.

The poison glands of the rattlers and other snakes of the pit viper group have been evolved from the salivary glands of the upper jaw and lie back of the eye. From them the venom, a yellowish, slightly viscid fluid, is conveyed to the fangs in the front of the mouth. These, true hypodermic needles, are hollow, sharply pointed teeth which serve as the actual instrument for injecting the venom into the body of prey or enemy. The fangs may be distinguished from the other teeth and from the teeth of harmless snakes not only by their greater size, but by the fact that they are not fixed and rigid, but are freely movable at the will of the snake. Ordinarily folded in a protective sheath of membrane against the roof of the mouth, in threat or strike they are swung down and out so as to project slightly from the opened mouth. The fangs do not necessarily move as one, but are capable of independent motion. It is not an unusual sight to see a caged rattler or copperhead open its mouth as if yawning and lazily erect first one fang and then the other.

The fangs are not permanent but are replaced by new ones periodically. Naturally, if the snake had to wait as other animals do for a new tooth to grow in place of a lost one, there would be a period in which he would have neither protection nor the power to kill his food. On each side of his upper jaw, there are two fang sockets instead of one. These alternate as bases for active fangs. The new fang has swung into the empty socket beside the one about to be discarded. After it is firmly set in place and its connection with the duct of the venom gland established, the old fang will loosen and be left in the body of the next prey. Behind the active fangs may be seen a series of smaller fangs, ranging in size from newly budded to almost full length, waiting their turn to move forward into service.

It is to be noted that the bite of one of these snakes is not a grab and bulldog chew as in the case of venomous snakes with fixed fangs, but is a quick stab which permits of lightning-like withdrawal and the minimum possibility of the snake becoming caught and injured by the bites and struggles of the prey.

We have proof of this stabbing speed in the many authentic records of cases in which an individual was bitten several times before he could get out of striking range. These reports of multiple bites are usually from the tropics and the snake the bushmaster, the fer-de-lance or other aggressive and vicious relatives of our northern pit vipers. Nevertheless, multiple bites are within the power and capability of the rattler, the copperhead and the moccasin and occasionally occur. Angry snakes, closely cornered, and captive snakes, deceptively tame and too casually handled, have been known to inflict several severe bites in quick succession.

For most snakes, the strike is a last resort and they can usually be depended upon to heed the warning noise of the approach of a human and slither quickly into hiding or tighten into a compact, inconspicuous resting coil. If unseen, and passed by—well and good. Discovery and aggressive behavior on the part of the intruder result more often than not in frantic attempts to get away. This is particularly true of the northern rattlers. A method not in the least intended by the snake is often the means of his making a clean escape. In the mind of the human assailant, the idea suddenly pops up that the snake is *after him,* and he reacts accordingly. The actions of the snake which give this impression are simply the result of his trying to do two things at once. He tries to get away to the side and to keep on the defensive—facing the intruder, enough slack in his upper third of body to give him striking power, tongue questioning the air. Impairing flight by attempted defense and defense by attempted flight, he is not efficient at either. Yet, giving as he does the impression of charging in a sly and devious fashion, he is often practically successful in both.

With the exception of the great diamond-backed rattlers and perhaps the water moccasin, there is no great likelihood of snakes striking above the knee. Boots or high shoes and heavy canvas or leather leggings afford good protection. If possible, the boots should not fit snugly about the leg. Loosely fitted materials with plenty of "give" are much more difficult to puncture than the same material stretched tight. Overalls or pants of a similar tough, closely woven material, worn on the outside of the boots and hanging well down over the ankle, give added security.

Cracks in old stone walls and foundations, the natural crevices of hilly country and the tumbled boulders of dry stream beds are often snake-infested. Many persons have been badly bitten in the hands while scrambling about such terrain. If it is necessary to use bare hands in climbing, a close lookout for snakes should be kept, lest the hand be quicker than the eye and a snake quicker than either.

Contrary to the accepted idea that the venom is intended as a means of offense and defense, its primary function is that of food-getting. Just as the careful hunter conserves his ammunition, so the rattler and his relatives conserve their venom. The injection of venom is not an automatic part of the strike but is a controlled phase of it. Mice and similar small game require only a small part of the contents of the venom sacs to bring about their quick death, and the snake is able to adjust the size of the dose to the need. Angry, injured or frightened snakes use no such discretion and may nearly empty the glands in a single strike, even, as sometimes happens, discharging the venom into the air in an unsuccessful strike.

The effectiveness of the venom, drop for drop, varies with the several species of pit vipers, but in composition they are all basically similar. Two main types of poison, each different in the symptoms produced, are combined in varying degrees to make up the venom. The first has a quick destructive effect upon the red blood cells and upon the tissues in the region of the bite. The power of this *hemotoxic* poison is immediately noted in the swelling around the wound, the discoloration of the parts and the accompanying intense burning pain. The *neurotoxic* element in the venom acts upon the nervous system and particularly upon those nerve centers controlling respiration and the heart.

Many factors besides that of the relative venomousness of the species concerned determine the seriousness of a snake bite. The age and weight of the patient (naturally the same quantity of venom will have a more profound effect upon a small child than upon a grown man); the mechanical effectiveness of the bite—direct or glancing, partly blocked by clothing or landing in bone; the site of the bite—in the body or limbs; in fat, which slows up its passage into the blood stream, or in a blood vessel, which carries it quickly to the heart. The size and condition of the snake brings in another group of varying determinants.

A most important factor in the ultimate severity of a venomous snake bite is the attitude and behavior of the patient and his companions. Excitement and emotional carrying-on speed up the circulation and give the poison a better chance to distribute through the system. The bitten person's duty to himself requires that if there is anyone around to do things, he should lean back and enjoy his chance at being the center of a lot of other people's activity.

Almost certainly, the first instinctive reaction of a person bitten is to suck the wound to rid it of poison. This, while a valuable and efficient means of first aid, is dangerous and should not be done unless other first aid is out of the question. Although snake venom is harmless when taken into the stomach, it can produce its characteristic symptoms as readily when introduced into the blood through split lips or sores within the mouth, as by a bite. The case of a boy working on a trail who was bitten in the leg by a rattler was widely publicized some years ago. A fellow worker sucked the wound after a second had opened it with his knife. The ambulance came and took the patient to a hospital for antivenin injections and, except for the inevitable discussions of the event, the matter was presumably closed. A little later the samaritan who had sucked the bite complained of dizziness and his lips became badly swollen. He had infected some tiny mouth lesion with venom by his first aid. The ambulance made another trip up the rocky road. Later in the day, the gentleman with the sharp pocket knife sliced his finger while cleaning the knife. Venom and blood still gummed the blade; his hand began to swell and he hysterically announced that he, too, was poisoned. The ambulance was used to the trip by that time. An hour later the three sheepish victims of a single bite had a chance to compare symptoms from adjoining beds. Such cases are rare, of course; and the newspaper accounts of this episode were no doubt subject to the usual exaggeration. Nevertheless, in giving first aid, precautions must be taken to avoid the possibility of secondary poisonings.

Before treating a snake bite as poisoned make sure the snake was actually of a venomous species. The bite of the rattlers, the copperhead or the moccasin consists of a pair—sometimes only one—of punctures, accompanied sometimes by lacerations from other teeth, which may resemble the jagged tears or scratches inflicted by our harmless snakes, except for the large fang marks. The presence of venom is further made known immediately by the intense, burning pain around the bite. Swelling and discoloration speedily follow. First aid begins with the placing of a tourniquet between the bite and the heart. Heavy string, a suspender strap or shoe lace, anything that will restrict the flow of venous blood back to the heart, will do. Next, use a sharp knife or razor blade, sterilized by a momentary holding in the flame of a match or by dipping in iodine or alcohol, to open each puncture to a depth equal to the fang's penetration. Too many cuts tend to isolate blocks of tissue from the blood supply. These become necrotic and add to the possibility of a bacterial infection developing later. A cross incision over each puncture or, in the bite of a smaller snake, a single lengthwise cut connecting the two, is sufficient. The next step is suction. This may be done by mouth or, less dangerously and more efficiently, by a suction cup of the type now commonly sold in snake bite treatment kits. These kits also include an instrument for opening the wound, a rubber tube tourniquet, iodine for sterilizing and other necessities. The suction cup consists of a metal nozzle (two types are supplied with each kit) attached to a soft bulb. With the bulb compressed, the nozzle is applied to the opened wound and the bulb released. The rounding back of the rubber bulb to its original form exerts enough suction to pull blood and venom from the bite. Suction should be continued until medical aid arrives. *The tourniquet must not be kept tight indefinitely*. Release it for a few seconds every ten or fifteen minutes. Complete stoppage of the blood over any period of time results in danger of gangrene in the parts.

Modern science has considerably mitigated the dangers and ill effects of snake bite, not only by providing specific treatment in the form of serums for injection, but also by bringing about the almost universal discard of the primitive methods used by the pioneers. It is generally accepted today that whisky or other alcoholics are of little value in such cases and, because of the effect upon the heart, likely to be harmful. The history of the settling of the West is full of stories, truthful as nearly as one can tell, of miners, cow-punchers, mule skinners and other swashbuckling citizens who, after being bitten by a rattler, consumed the raw and potent whisky of the frontier to the tune of a couple of quarts as treatment. The fact that they managed to survive to tell the story over and over despite the combined onslaught of snake venom and acute alcoholism is either a tribute to their toughness or a direct reflection on the effectiveness of the rattler's bite. There can be little doubt that these cases of easy recovery were the result of an encounter with the pigmy rattler, the sidewinder or the prairie rattler—all smaller species—or more likely still, were cases in which a non-venomous snake received credit for power he did not possess. The bite of either of the diamond-backs or of the timber rattler is not a light matter, and it is to be seriously questioned whether even men of the early West's fabled stamina could recover from such a bite without having received skilled and quick attention.

Antivenin is obtained by producing an immunity to the effect of snake venom in horses by means of periodical and gradually increased injections of the venom itself. When tests show them ready, these immunized horses are deprived of blood. The clear part of the blood, the serum, contains the immunizing element, and after having been sterilized and concentrated, is placed in sealed glass ampoules ready for use. Injected into the blood stream of a bitten person, it brings an active and specific agent into the fight.

If serum is available, administer according to the directions in the box. It comes with a hypodermic needle in a container and is conveniently ready for immediate use. Its value is not only in checking the effects of the venom upon the nerve centers but also in neutralizing the poison in the region of the bite and in this manner preventing the destruction of tissue.

No matter what is being done in the way of first aid, the services of a physician are imperative. Hospital treatment—if the patient can be taken to a hospital without too much of his own exertion—is also indicated.

Treat the wound with all care possible. Keep covered with a wet dressing of clean (sterile if possible) gauze soaked in a mild antiseptic. Snake venom destroys the power of the tissues to combat bacterial infections and these must be guarded against.

The supposed immunity of certain individuals and of certain animals to the effects of snake venom has evoked a great deal of heated argument in lay circles and has resulted in a large body of scientific discussion and experiment. Among animals, the hog has long been supposed to be immune to snake venom. It is a common saying in rural districts that the way to rid a tract of land of snakes (venomous or otherwise) is to pasture it to hogs for a season or so. Authentic case histories of porkers that have been bitten by rattlers or copperheads are lacking, but there seems to be adequate foundation for believing these animals to be immune, if the term is loosely construed. Their "immunity" is not physiological, but, one might say, mechanical. The heavy layer of fat blanketing the well-conditioned hog effectively holds the venom and dribbles it into the general circulation in such small quantities that its effects are lost or at least greatly impaired.

For years it has been supposed that venomous snakes were totally immune to the effects of their own venom or the venom of related species. The experience of most observers of captive specimens seems to bear out this theory. In feeding caged rattlers or moccasins, it often happens that one individual will seemingly strike another. Sometimes an angry snake accidentally inflicts a bite upon itself, with no apparent ill effect. Work on this phase of immunity, however, indicates that perhaps this is not necessarily true and that our idea of the immunity of venomous snakes has been due to faulty and incomplete observation and study.

The king snakes of the genus Lampropeltis have an actual immunity to the venom of the rattler, the copperhead and the moccasin. These strong constrictors suffer no apparent ill effect from the bites sustained in their frequent combats with venomous snakes, and their cannibalistic habits probably act as a substantial check on the numbers of the latter.

Huge fangs, poison glands of great capacity and the ability to inflict a deep and powerful bite through clothing or even leather, place the diamond-back rattlers among the first in rank of the venomous snakes of the world.

THE EASTERN DIAMOND-BACK RATTLER

Fortunately, the eastern diamond-back, *Crotalus adamanteus*, comes into comparatively little contact with man, due to its retiring habits and its choice of home. The wild swamps of the southeastern states, especially Florida and Georgia, unliked and little visited, are its common haunts and in them it is king. Attaining a recorded length of more than seven feet and the girth of a small car tire, the diamond-back is bold and its behavior in an emergency is cold, calculated and vicious. It is not aggressive, but in moments of uncertainty and danger quickly throws itself into a defensive, menacing coil, with a rasping noise of rough scale sliding on rough scale. This compact and tense bulk forms a solid base for the upper third of its body, draped in an S-shaped loop, to push against in strike. There is no evidence of excitement—no hysteria. The questioning tongue and the glitter of the jet-black eyes are the only evidence of life. Finding itself observed, the diamond-back sounds its rattle, a sudden, loud and rasping whir, filling the air, difficult to locate. Coming unexpectedly from beneath a clump of palmetto or (as John Orth and I once found it), from beneath the lustrous and fragrant bulk of a magnolia in full bloom, the sound is hard to locate and the hearer is faced with the urgent problem of standing still alongside grave possible danger or moving and stepping into or upon it. The pattern of yellow-edged diamond markings admirably simulates the haphazard hit-and-miss streaking of sun through grass and leaves and makes the snake inconspicuous. Once sighted, the bold, symmetrical markings, delicately graduated coloring and vigorous beauty of mass of the reptile combine in a picture of unforgettable magnificence and impressiveness.

The eastern diamond-back is irritable, surly in temper and not particularly adaptable to captivity. Caged, it is almost continually on the lookout for trouble and spends most of its time in a defensive coil, rattling at any disturbance, fretting itself out of life. In nature, it feeds largely upon rabbits. In captivity, some individuals will feed, if let alone and secure from interruption, while others eventually starve to death unless forcibly fed—a most dangerous proceeding for the amateur. Diamond-backs born in captivity—the young are usually about a foot long at birth and come in litters of as many as twelve—feed readily upon mice and have in a number of cases been successfully brought to maturity. Such snakes are less irritable, tamer and more satisfactory for exhibition than captured snakes.

THE TEXAS DIAMOND-BACK

The western diamond-back, *Crotalus atrox,* is almost as large as its eastern relative, equally irritable and surly, but more amenable to captivity and less inclined to suicidal starvation. Its colors are weaker and its patterns lack the symmetry of

U.S. Fish and Wildlife Service, photo by Francis M. Uhler

CLOSE-UP OF MOUTH OF TEXAS DIAMOND-BACK RATTLESNAKE

C. M. Bogert

TEXAS DIAMOND-BACK RATTLESNAKE

PRAIRIE RATTLER

TIMBER RATTLER

PIGMY RATTLER (MASSASAUGA)

the beautiful eastern diamond-back, although occasional specimens are vivid and rich in pattern. The western diamond-back
and its subspecies have a wide distribution throughout most of
the Southwest and are responsible for more fatalities than any
other of our American serpents. Their seeming abundance in
the semi-arid regions of Texas, particularly, has brought them
into contact with man, with no pleasing result to either side.
In nature, they feed largely upon small warm-blooded game of
the desert: cottontail rabbits, mice, kangaroo rats, gophers.
In captivity, their tastes run along these lines. Because of its
numbers and collectibility and its great size, the western diamond-back is the diamond-back of both the traveling snake
show and the zoological collection.

THE SIDEWINDER

The southwestern states possess a greater concentration of
rattlers, in numbers as well as in species, than other regions.
The semi-arid plains and hills offer an abundance of small,
warm-blooded creatures, which is reflected in the variety of
species and habits of the snakes. Of the smaller and more unusual rattlers of the Southwest, the "sidewinder," *Crotalus
cerastes*, is the most curious. A dweller in sandy, desert regions,
the sidewinder has developed a manner of locomotion similar
to the looping movements of the true vipers of the African
deserts. This movement carries the snake over soft, yielding
sand with the maximum efficiency and leaves a distinctive and,
at first sight, puzzling track. The sidewinder is often known as
the horned rattler because of the enlarged scales which jut
out above the eyes. These "horns" are an easy distinguishing
mark.

THE PRAIRIE RATTLER

The rattler of widest distribution is the one commonly
known as the prairie rattler, *Crotalus viridis*. Slimmer and
smaller than the diamond-backs, averaging no more than
three or three and a half feet in length, although specimens as
long as six feet have been recorded, it is widely distributed
west of the Mississippi, down into the northern parts of Texas,
north into Canada and, in its subspecies, *oreganus,* the Pacific
rattler, up and down the west coast. In nature, the prairie and
Pacific rattlers are savage in behavior, and resentful of human
interference. They put up a vicious battle against capture and

strike repeatedly in the direction of their assailant. Once in captivity, they tame readily and, with the exception of a few individuals, settle down to a contented cage life. The important thing is to keep in mind the fact that they are highly dangerous and not to be treated casually. Without realizing it, owners of these snakes often treat them with an extraordinary lack of respect—handling with bare hands, putting water dishes into the cage while snakes are within striking distance, and so on. Such practices are foolhardy and not to be countenanced.

The usual color scheme of the prairie rattlers is some shade of delicate olive-green or yellowish green, with darker light-rimmed blotches along the back. The head is not of an ugly brutish type but is delicately formed and marked. The Pacific subspecies is found in a considerable variety of terrain—high mountains and river bottoms, flat valley land and bald hills—and naturally presents a number of slightly different habitat color phases.

The greatest problem in feeding the prairie rattlers is to get enough mice and small rats for them. They develop riotous appetites, especially when kept in large groups, and can consume an enormous number of mice weekly. The writer successfully conditioned a number out of a large group to eat strips of lean beef and beef heart.

Although no rattler can be described as arboreal, the prairie species seems to be more comfortable if the cage contains a few shrubby bushes to crawl around in when the notion strikes.

The hunting habits of the prairie rattler have been responsible for one of the most celebrated snake legends—that of the prairie dog, the burrowing owl and the rattler all living in amity and friendship in the same hole. There isn't the slightest doubt that the rattler *is* often found in the burrows of the prairie dog, but it is hardly companionship he is after. He is no honored house guest but merely an uninvited and unwelcome guest at dinner—a dinner composed of the helpless young of the proprietor of the establishment.

Early annals of America abound in stories of enormous congregations of rattlesnakes assembling in certain localities to hibernate for the winter. This habit is especially characteristic of the prairie rattlers and the timber rattlers, but the old-timers' stories of balls of snakes as large as hogsheads being uncovered seem a trifle excessive.

THE TIMBER RATTLER

The timber rattler of the eastern states, *Crotalus horridus,* and its habits of hibernation have inspired most of the "tall" tales. Never found very far from quick access to some stony crevice or crack, this species has certain favored hibernating sites and gathers in these places in considerable numbers in the fall. The opinion that the selection of a den site is governed by ancestral tradition and that snakes return to the same hibernating den year after year is held by some herpetologists. Others believe that a warm exposure, adequate drainage and a terrain sufficiently broken to permit the snakes to find deep and safe hiding places are the factors which make a site suitable for a snake den. Even a casual survey of a stretch of timberland will show that surprisingly few perfect den locations exist. In a region in which rattlers and copperheads abound such locations almost always have a considerable local reputation.

Sunny days of early spring call the snakes out from their winter sleep and, for a week or so, during which courtship and mating occur, they may be found in large numbers, basking on ledges. Later, as the weather warms and danger of freezing passes, they scatter out through the woods in search of good feeding grounds.

The pattern of the timber rattler is composed of alternate, somewhat irregular light and dark bands which tend to obscure as the skin becomes old and the heavy keeling of the scales fills with dust and grime. When freshly shed, the bandings are bright and the rattler is one of the most beautiful of snakes. The ground color in those of the East is usually olive or a dull rich yellow with bandings of velvety, sooty black or dark brown. The bands are edged with a scale's width of brighter color and the tail is often entirely soot brown or black. This rattler is subject to a great deal of variation throughout its range and there are distinctive color phases peculiar to certain localities. Northern forms tend toward black with an obscuring of the bands. In the South the rattlers of this species are larger in size and brighter in color. A southern subspecies, *Crotalus horridus atricaudatus,* occurring in the lowlands and swamps from North Carolina to Florida and across to Louisiana, is known from its habitat as the cane brake rattler. One summer in the heavy-aired and sunless depths of Black Swamp in South Carolina, we came upon a large five-foot female loosely coiled but defensive, glowing in new beauty of delicate salmon pink barred with a bold and vigorous design of angular black bandings.

These southerners are more vicious and less amenable to the restrictions of cage life than northern varieties of timber rattler. The latter compare favorably with the prairie rattler in docility and the ease with which they adjust themselves to captivity. In nature the timber rattler preys upon warm-blooded small game and birds. Its successful care means that a constant source of food in the form of mice or young rats must be arranged for or maintained.

THE PIGMY RATTLERS

The only rattlers noted for taking cold-blooded food are the pigmy rattlers of the genus Sistrurus. The southern species, *Sistrurus miliarius*, found throughout the flat lands of Florida and the Carolinas and westward along the coastal plains to Texas, is the smallest of the rattlesnakes and attains an adult length of but sixteen or seventeen inches. It is smaller than its northern cousin, the massasauga, *S. catenatus catenatus*, or the western species, *S. c. tergeminus*, found as far west as Arizona. Both the southern pigmy and the massasauga favor damp and swampy habitats and feed upon frogs as well as mice, small birds and such game.

This choice of habitat and food as well as the large shields of the head indicate a close relationship to the copperhead and the moccasin.

THE COPPERHEAD

Although the scalation of the heads of the copperheads and the moccasin is like the scaling of many harmless snakes in that the head is covered with shields rather than by small granular scales, they may be easily recognized by the pit between the nostril and eye. Too, the elliptical pupil of the eye, in shape like that of a cat, is markedly different from the round pupil seen in the eyes of our non-venomous snakes.

Although vigorous in defense if it believes itself in danger, and capable of inflicting a severe and potent bite, the copperhead, *Ancistrodon contortrix*, is shy and retiring and tries to avoid any encounter with man. Its coloring is markedly beautiful, especially in a newly shed specimen, and admirably adapted to rendering the reptile inconspicuous in the welter of fallen leaves and debris of the forest floor. Although subject to a great deal of variation, a rich brown with an overcast of dark coppery red is the usual phase. The head particularly is shaded with the coppery, metallic tint which has given the snake its popular name. A pattern of saddle shaped bands of a darker and more velvety character than the ground color camouflages the snake and gives it another colloquial name, the poplar leaf snake.

The copperhead has been somewhat unfairly characterized in literature and folk lore. In reality, it is no more vicious and aggressive than its kin, the rattler, and because of its relatively small fangs less dangerous; nevertheless, it has become the personification of treachery, lurking danger and death.

Under ordinary circumstances the copperhead, if disturbed, will quietly and unobtrusively attempt an escape. If cornered, defense is vigorous and even hysterical. Unlike the larger rattlers, it does not reserve its strength and venom for a telling and lethal strike but strikes wildly in the direction of its enemy. Instead of resting in a compact and protected coil, its body may toss about in a threshing series of loops and the tail beat a frenzied and buzzing tattoo among the sticks and dry leaves. These frantic tactics are the cover-up for a quick retreat to the shelter of the undergrowth or to some convenient crack or crevice of the rock.

The range of the copperhead is throughout the eastern, southern and central states as far west as Oklahoma and Texas. In the North, its haunts are the same as those of the timber rattlers, heavily timbered and rocky regions, and it has been found to hibernate under the same conditions and in the same dens. Unlike the timber rattler, it is to be found also in the damper parts of the woods or along the borders of marshes and streams. This is because of its fondness for frogs, an article not on the rattler's diet list.

The copperhead of Oklahoma and Texas differs from that of the East in having wider markings and a generally darker coloring. Another difference is that the tail retains in maturity some of the bright yellow coloring so characteristic of the tails of baby copperheads.

In captivity, after a few watchful and perhaps restless days, the copperhead usually settles down to a quiet and contented life, and makes an attractive, well-behaved, but highly dangerous exhibit. Its food preferences are somewhat captious and variable and must be indulged if the creature is to be kept in health. Feeding through the year on a varied diet of small rodents, birds, frogs and small snakes, a copperhead apparently well-conditioned to a diet of frogs will suddenly refuse its accustomed food and enter into a fast which can only be broken by a meal of rodent or bird.

Like its kin, the rattlers, the copperhead gives birth to a small brood of living young—usually from six to nine in number—in the late summer or early fall. The young are scrappy from the start and able to inflict a bite sufficiently venomous to make the victim quite uncomfortable.

THE WATER MOCCASIN

Through the South the favorite name for the dreaded water moccasin, *Ancistrodon piscivorus,* is the cottonmouth. This refers to the whitish appearance of the inside of the snake's mouth and its habit, when startled, of quickly assuming a striking position, and with widely distended jaws and erect fangs threatening the intruder. The lowlands and swamps of the Carolinas and Georgia and the famed Everglades of Florida are the home of this unattractive, bad-tempered and venomous serpent. Stocky and ungraceful in movement, with a squat and ugly blackish head, sharply distinct from the slender neck and with an abruptly tapering tail, large moccasins are hardly beautiful. Newly shed specimens of not too great a size, have a pattern quite similar to that of the copperhead and a coloring equally showy, although the prevailing color runs more to olive-green than to copper-reds. Indeed, the young of the moccasin often resemble the young of the copperhead in color and pattern so closely as to be confused with them.

In captivity, moccasins are easy to care for and adjust readily to a caged life. Feeding in nature impartially upon small warm-blooded game, fish, frogs and other species of snake, they take with gusto almost whatever the keeper offers. If possible, their cage should have a water container large enough for the snakes to submerge themselves completely. Such a tank offers not only a substitute for the ponds and marsh water so integral a part of the serpent's natural environment, but gives the keeper a chance to educate his snakes to a cheap and easy method of feeding. (See Ribbon snakes, page 154.) Very active at the time of capture and aggressive for some days after, moccasins deceptively settle down in captivity into well-behaved and gentle creatures seemingly oblivious to everything except food and downright abuse. They show no hostility,

U.S. Fish and Wildlife Service, photo by S. E. Piper

PACIFIC RATTLER WITH PARTIALLY SHED SKIN

U.S. Fish and Wildlife Service, photo by E. A. Goldman

WATER MOCCASIN, SHOWING "COTTONMOUTH"

SONORAN CORAL SNAKE

COPPERHEAD

and it is only when they yawn that a glimpse is caught of the menacing white "cottonmouth." There are no obvious signs, but the potential danger should not be overlooked or over-shadowed by the carelessness bred of familiarity. The moc-casin's fangs are long, its venom powerful, its strike sudden and without warning.

THE CORAL SNAKE

In addition to its share of the pit vipers, the southern United States has representatives of another group of venom-ous snakes, the Elapidae. Members of this group in other parts of the world attain large size and exist in such numbers and venomousness as to be a formidable menace. Examples are the Asiatic and African cobras. American Elapidae, belonging to the genera Micrurus and Micruroides, called by us the coral snakes, are small and somewhat degenerate types of their group. Burrowing in habit and seldom coming to the sight of man, these snakes, although structurally like the cobra and the mamba and with a similar venom, hardly constitute a vital danger. This relative harmlessness is *not* an attribute of the individual snake. Drop for drop, the coral snake venom is much more effective than that of the rattlers or even, accord-ing to some experts, than that of the cobra of India. The small, immovable fangs, too small to pierce shoe leather, and a sluggish reluctance to bite, have saved many persons unpleasant results from an encounter with one of them. The few authentic records bring out the fact that when a coral snake bite occurs, the case is a serious one.

The great danger from coral snakes lies in confusing them with the similarly patterned and colored harmless king snakes found in the same region. Rings of bright coral red and glosssy black are, in the coral snakes, separated by bands of rich yel-low. In the harmless king snake, a slightly different arrange-ment of the colors furnishes a certain identifying difference— the yellow and red are separated by the bands of black. In the coral snakes, the bands of color encircle the body and appear on the belly surface much as above, while in the similarly col-ored harmless species, the bands are broken and the ventral surfaces either whitish or lightly blotched.

The coral snakes of the United States are seldom more than thirty inches in length, slim, blunt-tailed, almost cylindrical in shape and with a rounded blunt head admirably adapted for burrowing. Small snakes and lizards, the blue-tailed skink of the South particularly, are its food, and it is often found in old wood debris, under rotten logs and in similar locations where these lizards abound.

In captivity, the coral snake requires a temperature approximating that of its southern home, living food in the form of lizards or small snakes and a cage filled with peat or sphagnum moss, rotted wood, and earth in which it may conceal itself

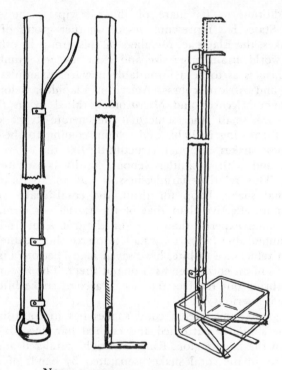

NOOSE STICK, HOOK AND WATER LIFTER

effectually and permanently from view. Although brilliant in color, its erratic food habits and its venomous quality, coupled with the fact that it can never be seen if properly kept, make it emphatically *not* an advisable guest for the small zoo.

Cages for venomous snakes are essentially the same as for non-venomous with the exception that all wire screening should be double, the two layers being an inch apart. Venomous snakes occasionally strike through wire and have been known to deliver through single mesh a serious bite which a second layer would have prevented.

In transferring rattlers from cage to cage, there is little need for handling. A hook made of a three-foot stick with a metal right angle screwed to the end is perfectly safe and convenient. If medical treatment is needed or for some other reason it is necessary to handle a venomous snake, the use of the noose stick illustrated reduces danger to a minimum.

Turtles

THE PRINCIPAL characteristic which distinguishes the turtles from other reptiles is the possession of a body shell into which soft and vulnerable parts—head, legs and tail—may be completely or partially withdrawn. Some species, such as the gentle and inoffensive box turtle, a land dweller, have the ability to retreat entirely within the protective shell and close it after. Pugnacious and vicious turtles like the snappers are unable completely to protect themselves in this curious fashion and have to depend on the more common defense of snapping and biting.

Except in the case of the soft-shelled turtles, the upper part of the shell, the carapace, is made up of the modified and fused ribs and vertebrae, with an additional number of bony plates. These are covered with a thin sheathing of horny substance. Fastened to the carapace and with it forming the complete turtle shell, is the plastron, the lower section, hinged in some species and in others solid and unyielding.

This rigid construction renders the body inflexible and the turtle is able to move only its head, neck, legs and tail. Even the breathing is affected. No bellows action of the lungs is possible within this tight cage, and one way turtles get air into the lungs is by swallowing. Startled turtles, pulling in legs and head, usually give vent to a short hiss. This noise, rather than being an expression of fear or excitement, is the sound of breath being expelled to make room for parts in need of protection.

Turtles may be divided roughly into two groups: those living for the most part in water and those living on land. Although structurally much the same, the two groups have certain obvious differences. The water turtles have more or less webbed hind feet for swimming and their shells are fairly smooth and flattened. The land turtles and tortoises, generally speaking, have feet adapted more for walking than swimming and their shells, instead of being flattened, are domelike and high.

THE LARGER TERRAPINS

Central and eastern North America are favored with a great many species of fresh-water turtles. With the exception of the snapping turtles, the mud and musk turtles and the soft-shelled turtles, these are usually grouped under the name of "terrapin," a term which has little scientific standing, but which does mass conveniently a number of species of similar habit and appearance under one heading. The larger terrapins are edible—the famous diamond-back is the gourmet's idea of a heavenly dish—and through their common use for food the term has taken on a shade of meaning implying edibility.

The larger terrapins are found in ponds and lakes, streams and rivers and, as in the case of the diamond-back, even in the eastern coast salt marshes lying along the sea.

Although the water is their real home, their protection and their hunting ground, the terrapins have a fondness for the upper air. Floating logs, half-submerged rocks, small islands and even the less safe stretches of sandy pond or stream side, offer highly desirable sun-bathing facilities. Often they may be seen atop some meager roost in the sun, piled one on top of another, scrambling and clawing to keep from slipping back into the water. The quick scuttle and plop with which they take to the water when disturbed has given them the popular name of "sliders."

At egg-laying time, terrapins venture farther on land than at any other time. In a suitable spot, each female scoops a hole in the earth or sand with her hind legs. Her method of digging is not a mere haphazard scrabbling at the ground; each hind foot slowly and methodically brings up its load of sand in turn and dumps it to the side. The eggs, varying in number and size with the species, are laid in the hole and covered by a few scooping movements of the hind legs. The female returns to the water, leaving the eggs to the good offices of time and the sun and to the mercy of such canny hunters as the raccoon and the skunk.

The incubation time of the eggs is variable, but in due course, if some of the aforesaid hunters have not found them, the hatching young slash a way out of the enveloping shell by means of an egg-tooth; then they emerge and struggle up through the protecting blanket of sand. Guided and attracted by the light of the open sky above the nearest water, they find in the weediness of its shallows some measure of protection.

The youngsters of the larger terrapins are more suitable for the vivarium than their bulky elders. They have become popular as pets, and a pet shop tank containing several hundred hatchlings is no rare sight. Unfortunately, in an attempt to make these already attractive and beautiful little reptiles more salable, some bright lad conceived the idea of using paint to ornament the shells. Burdened with pink roses, landscapes and birthday greetings, the poor creatures are bought carelessly and casually in exactly the same spirit as a greeting card or piece of pottery. In this way, they find themselves in all sorts of surroundings, mostly unsuitable, and few of them long survive the excessive handling and neglect which is their ordinary fortune.

If a "decorated" individual should fall into your hands, do something about the art work for the sake of the proper growth of the turtle's shell and for the preservation of your own reputation for good taste. The paint ordinarily used is one of the quick-drying lacquers. For its removal, either the commercial lacquer thinner or pure acetone is necessary. Holding the turtle upside down, press a cotton pad saturated with the liquid against the paint job. The lacquer thinner is somewhat irritating (it is wise to wear rubber gloves) and care must be taken to apply it only to the painted shell. A stiff paper collar, held in place by pieces of adhesive tape, protects the turtle's head and eyes, not only from the fluid, but to some extent from the fumes as well. Some rubbing and perhaps several applications of the thinner may be needed before the unnatural ornamentation is destroyed and the turtle is a self-respecting animal again.

The terrapins whose infants most commonly suffer this decorative indignity belong to several very common central and southern forms. Of these the genus Pseudemys is the most important, including, as it does, the largest number of species and those which attain the greatest size. Much variation in color occurs within the species comprising this genus and has made their proper classification a difficult matter and their relationships confusing.

For many years two apparently distinct species of the Mississippi Valley were generally accepted as valid. These were the form known popularly as the Cumberland turtle, *Pseudemys elegans,* and a darker terrapin, *Pseudemys troostii,* considered distinct and easy of identification because of its lack of yellow markings on the head. In the course of years, hundreds of specimens of terrapins, young and old, passed through the

U.S. Fish and Wildlife Service, photo by E. P. Haddon

ALBINO DIAMOND-BACK TERRAPIN

National Park Service photo by Allan First

WOOD TURTLE

Robert Snedigar

PAINTED TURTLE

Robert Snedigar

BOX TURTLE

hands of a New Orleans herpetologist, Mr. Percy Viosca, Jr. Among these he noted many hatchlings and immature individuals which he could easily identify as *Pseudemys elegans,* but none which could be classed as young of the species *Pseudemys troostii.* Further examinations and measurements of several large series brought out the truth. Individuals of the dark form known as *troostii* were always large and male. Smaller and not fully mature males and *all* females had the patterns and coloring of *elegans.* In some cases males were found passing over from the *elegans* type to the *troostii.* What had been so long thought of as a difference in species coloring turned out to be simply a sexual difference. Now, instead of two species, we have one, *Pseudemys scripta troostii.*

Most of the Pseudemys have an undersurface of yellow, or of yellow with black markings. An exception is the red-bellied terrapin of the eastern states, *Pseudemys rubriventris,* in which the plastron is dull red or reddish-orange.

In captivity, the larger terrapins are hardy and, given adequate sun, thrive upon a diet of fish, chopped meat, and succulent greens such as lettuce or natural pond greens.

The terrapins of the genus Graptemys are markedly similar in size and general appearance to those of the preceding group. Like them, Graptemys are a common market terrapin and are caught in considerable numbers every year for food sale. Two principal species make up the genus. The first, called the map turtle or the geographic terrapin, because of the curious resemblance of the fine irregular lines on the shell to those of a map, grows to a considerable size. Full-grown females are larger than males and often grow to be as much as a foot in length. The carapace, like that of the other large terrapins, is solidly united to the plastron, and with it forms a rigid, inflexible case.

The geographic turtle is found in abundance through the Mississippi Valley and even ranges northward into Canada.

LeSeur's terrapin, *Graptemys pseudogeographica,* and its several subspecies share partly the same range as the map turtle, and resemble it in size and general character. As is the case with the other large terrapins they are marketed as food when adult and as pets when young.

In captivity, Graptemys have not proven as satisfactory as some of the other terrapins.

THE DIAMOND-BACK TERRAPIN

The delicacy of flavor of the diamond-back terrapin, *Mala-clemys centrata,* has been largely responsible for making the word terrapin synonymous with good eating. Unfortunately, this food value, coupled with water pollution, has seriously damaged the species and eliminated it from many parts of the range in which it was formerly abundant. Conservation measures on the part of the government, and protection and systematic breeding for market under semi-captive conditions have done much to prevent an irreparable loss.

Unlike the other terrapins, the diamond-backs are never found inland but prefer the brackish marshes along the ocean where an abundance of small creatures of the sea furnishes them with an ample food supply.

The most desirable commercially are the large females of the Atlantic coast. These, ordinarily seven and one-half or eight inches in length of shell, and weighing but a few pounds, fetch a very high price in the market. The males seldom grow to more than five inches.

In appearance and behavior, the diamond-back is the most attractive of all the water turtles. The coloring, although variable, is always clear and clean-looking. The carapace of a large female is usually of a light olive-gray, with slightly darker concentric markings within each shield and on the borders. The head, neck and feet are of a clean slate-blue, very light and finely speckled with black. The plastron is of a yellowish cast, more or less blotched with black. Males run darker in color; the carapace inclines to black and each shield is sculptured in a series of concentric rising ridges. These are the markings to which the name diamond-back refers.

In captivity, the diamond-back needs a sunny, warm tank with plenty of swimming room. Salt water, although desirable, is not essential. But it is an undoubted aid in the control and cure of the fungus disease of the shell to which turtles are susceptible.

Diamond-backs are, for reptiles, very intelligent and evidence their superiority in many amusing ways. Turtles learn quickly to associate human beings with food supply and to expect something to eat whenever a person comes near. In the case of the mud and musk turtles, anything put in the tank is snapped at first and investigated afterward. This includes fingers. The diamond-back, on the contrary, expecting food, paddles furiously or lazily—depending on how hungry it is—up to the object, investigates it and, if the decision is favorable, devours it. Fingers are nosed, but rejected. In captivity the diamond-back is conditioned without difficulty to easily obtained and cheap foods such as beef heart, liver and fish.

THE PAINTED TURTLE

With the possible exception of the Pacific Coast terrapin, *Clemmys marmorata,* the smaller terrapins are not generally considered edible. However, their smaller size and hardiness makes them more desirable for the small animal collection or for a vivarium pet. In habit, they are much like the larger species.

Of these smaller terrapins, the painted turtles of the genus Chrysemys are perhaps the most conspicuous in coloring and take their name from the vivid red markings of the shields bordering the carapace. In *Chrysemys picta picta* of the eastern states, the red is strongly repeated in the striping of the legs and the head is streaked with bright yellow. The carapace is an olive-brown, with each shield defined by a narrow edge of greenish gray or yellow. In larger specimens, of seven- or eight-inch shell length, the colors tend to darken and the markings become less distinct. The newly-hatched and young of this group display the pattern most effectively and well deserve the name of painted.

There is not a great deal of obvious difference between the sexes in the painted turtles, but mature males have a distinguishing character, easily seen, in the very long claws of the front feet.

Like most of the terrapins, the painted turtle eats practically anything in the way of small animal life that comes its way, together with a judicious amount of green plant material.

THE SPOTTED TURTLE

The spotted turtle of the eastern states *Clemmys guttata,* is not entirely aquatic. Its legs are less extensively webbed than those of the larger terrapins and it is often encountered bungling along through the dry leaves of the woods or in the wet grasses of swampy ground. Like the painted turtles, too, it basks in large groups, neck stretched out and legs sprawled to get the sun. A common haunt is in the tortuous channels threading salt marshes adjacent to the sea. Sculling along the bottom, it takes quick advantage of the overhanging grasses for cover when disturbed.

This turtle is somber in color. The carapace is of dull black, relieved by a polka dot, irregular pattern of yellow-orange spots. The number and location of these spots, as well as the intensity of their coloring, is subject to considerable variation even in a group of specimens caught in the same locality.

It was formerly believed that these creatures are unable to feed except under water—that a peculiar structure of the throat made swallowing impossible under other conditions. Although normally feeding in the water in nature, in captivity these turtles soon become accustomed to the presence of people and an ample supply of food. Nervousness and the dread of being robbed leave them; they become less wary of their fellows and, instead of dashing pell-mell for deep water where they can turn and twist, devour a land-caught morsel on the spot.

THE WESTERN TURTLE

Although rich in most forms of animal life, the Pacific slope is endowed with only one widely distributed fresh-water turtle. Close kin to the spotted turtle of the East, this western species, *Clemmys marmorata,* is somewhat similar in general shape and appearance, but attains a greater size. Specimens with eight-inch shells are often seen in the markets of San Francisco's Chinatown. In color, the western turtle is dark brown or brownish black, and the shields of the carapace are marked with yellow. The plastron is yellow. Head, tail and legs are brownish, blotched with black and sometimes yellow. When compared to the painted turtles and the diamond-back terrapin, the western turtle is at a decided disadvantage. It is not a showy animal and its principal interest is in the possession, undisputed by other turtles, of the territory west of the Cascade and Sierra Nevada Mountains.

In habit, these turtles are shy and, although as fond of sun-bathing as any terrapins, most apprehensive of danger and ready to slide into deep water at the slightest alarm. They are plentiful in the sloughs of the San Joaquin River and sometimes seriously interfere with the pleasure of fishing. Commercially, they find a ready market in the larger cities and are the mainstay and support of a fraternity of trappers.

THE WOOD TURTLE

The wood turtle, *Clemmys insculpta*, carries the terrestrial inclinations of its close relative, the spotted turtle, still further. Although ordinarily found on land, it has not entirely deserted the water and is a good swimmer. Without any great beauty of coloring or marking, this turtle is one of the most attractive of all. Instead of being smooth and even, the carapace has a roughly carved, sculptured look.

Its general coloring is a dull brown with faint yellow markings. The plastron is of yellow with a black mark on each shield, a marking that is repeated on the turned-over edges of the marginal shields of the carapace. The only brightness of color is in the brick red of the neck and the underparts of the tail and legs. The wood turtle is commonly found in damp woods and, especially in the spring, in marshy places. Wild strawberry patches are a favorite haunt when the fruit is ripe. Their food is made up of insects and their larvae, small fruits and tender plant material. Berries of all sorts are their weakness and captive specimens on a hunger strike have been persuaded by means of a luscious strawberry or two that life was after all worth while.

The range of the wood turtle is not extensive. It is apparently confined to the states of the central and northern Atlantic seaboard, although it has been recorded from as far west as Iowa and Wisconsin.

The wood turtle seldom attempts to snap and after a little time for adjustment, readily learns to take food from the owner's hand. In the fall, as colder nights and days of less sun approach, it refuses food and prepares for a winter nap. At this time, unless the turtle can be kept in sunny and warm quarters with adequate fresh vegetable food, it is advisable to provide proper facilities for hibernation.

THE SNAPPING TURTLE

In addition to the great number of terrapins and pond turtles, North America harbors one family of very large, pugnacious and ugly chelonians, the snappers. The most familiar species of the family, the common snapping turtle, *Chelydra serpentina*, has a wide distribution, being found from the Rocky Mountains to the Atlantic and from the Gulf of Mexico on up into southern Canada. Large and ancient individuals that have had the advantage of plenty of room and food sometimes weigh in excess of one hundred pounds. Those usually encountered are considered large if they weigh a fourth of that. Squat and heavy, enclosed in a rough shell inadequate for complete protection, the snapper depends for defense and food-getting on the darting thrust of the snake-like neck and the sharpness and strength of its heavy jaws. The head is massive and threatening and both upper and lower jaws are equipped with sharp cutting edges of horn-covered bone.

In color the snappers are a muddy dark brown, like the mud of the pond and river bottoms along which they glide in search of prey. In the summer larger snappers sport a rich pond-bottom coat of greenish algae on their shells which serves as a further disguise. The shields of the carapace of young specimens are heavily ridged, but with age there is a tendency for the keels to flatten out and the shell to become smoother. A conspicuous character is the sharp saw-tooth edge of the rear of the shell. Young specimens have very large heads in comparison to the size of the body, but, with growth, the ratio of head-size to body changes to a more conventional proportion.

The plastron of the snapper is unlike that of the terrapins or land turtles. Instead of being a fairly continuous stretch of protective shell, it is nothing more than a narrow supporting buckler and gives scant protection to the soft parts of neck, legs and tail. The feet are webbed for swimming and are equipped with heavy strong claws for use in tearing to pieces prey held in the jaws. The tail is longer than in most turtles and is the snapper's vulnerable spot, so far as man is concerned. Held up by the tail, the turtle is unable to reach the hand of the captor with its jaws and can only claw at empty air. In this fashion many a snapper has been taken out of circulation. If the turtle is not held away from the captor's legs, rage may be vented on them. Once with a good hold, the snapper is extremely reluctant to let go. A common superstition of the South is that once Mr. Snapper gets his jaws locked securely on something, he won't turn loose until it thunders.

The food of the snapper is almost anything in the way of small animal life that he encounters. Fish, gliding past the unmoving bulk, fall victim to the swiftly outthrust head and snapping mandibles. Wildfowl, especially young ducks, are dragged beneath the surface of the water and devoured. Frogs,

SNAPPING TURTLES EMERGING FROM NEST

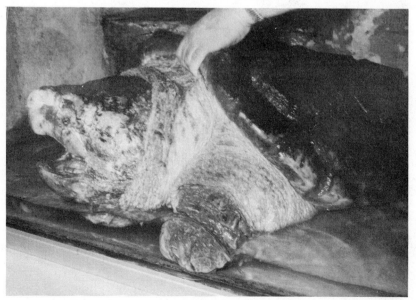

ALLIGATOR SNAPPING TURTLE, 236 POUNDS, PRIZE SPECIMEN
IN CHICAGO ZOOLOGICAL PARK

DESERT TORTOISE

unaware of the approach of the silent and dun-colored enemy, are taken by surprise and gulped.

In captivity, snappers retain most of their natural viciousness and if large should be handled with respect and caution. A good test of the strength and sharpness of its jaws and their possible effect on a human finger may be had by trying the turtle out on a stick.

The alligator snapper of the lower Mississippi and the Gulf states, *Macrochelys temminckii,* is larger than the common snapper, but much like it in habit and general appearance. Equally unpleasant in disposition, heavier-jawed and generally more obnoxious because of greater bulk, the alligator snapper is vouchsafed an aid in obtaining food which is not given other turtles. He is a skillful fisherman and adept in the use of artificial bait, a bait which is one of the most curious and unusual adaptations found in the reptile world. Lying quietly upon the muddy river bottom, partly obscured by the murkiness of silt-laden waters, the alligator snapper opens his jaws and the trap is set. Within, conspicuous against dun-colored mouth parts, floats a pinkish appendage, a tag of flesh whose movement and appearance call out to passing fish that here is a worm. To get the counterfeit worm, the fish must pass within the gape of the great, knife-edged jaws. The fatness and size of the average alligator snapper indicate how well the trick works.

THE MUSK TURTLE

The musk turtle, *Sternotherus odoratus,* and its relatives often exist in ponds and streams in considerable numbers without giving the casual observer any indication of their presence. Completely aquatic except for the female's annual egg-laying jaunt, they have no great fondness for the sun and light of day. Their color is a dull and muddy brown, often partly concealed by a thin overlay of green algae.

The carapace is smooth, oval and rather highly rounded. It is small, seldom measuring more than three or four inches. The plastron is narrow and, like that of the snapping turtle, affords but incomplete protection to the softer parts of the legs, neck and tail. The head tapers to a narrow, hooked beak and is adorned by two lines of yellow which start at the tip of the snout and run above and below the eye.

Sexual differences in the mud turtles are not distinct, but

males may be identified by their longer tails and by the patch of rough scales found in the bend between thigh and leg.

In some localities, these turtles are called "stinkpots," because of the disagreeable, musky odor emitted by special glands.

In disposition, they are vicious. Their unpleasant temper, coupled with their coloring and appearance, has resulted in their being referred to as miniature snappers. Equally ready to snap and bite, their attack is quite unlike the snapper's quick and forceful thrust of head and body. Slyly stretching out its neck to the utmost and so unable to put any force of body behind its bite, the musk is usually able to do no more than attempt an ineffectual nip.

THE MUD TURTLE

The mud turtles of the genus Kinosternon are somewhat like the musk turtles in superficial appearance. The most obvious difference between the two genera is in the plastron. That of the musk turtle is narrow and rigid while the mud turtle's plastron is efficiently hinged and is wide enough to effect a good protective closure. The mud turtles are very like the musk in behavior and food habits but less consistently aquatic.

In captivity, the musk and mud turtles thrive under adverse conditions but, lacking both attractive appearance and interesting behavior, haven't a great deal to recommend them. Often they are added to the small animal collection almost without the owner's being aware of it. Fishermen often find, after a fight with what was presumably a big fish, that they have hooked one of these bottom prowlers; having to dispose of it somehow, they are prone to deliver the unwanted captive to the nearest camp museum or zoo.

THE BOX TURTLE

In addition to its large assortment of water turtles and terrapins, the United States has its quota of terrestrial turtles and tortoises. Originally referring only to the true tortoises, the latter term has become, through general usage, applicable also to a genus of common land turtles, Terrepene, the box turtles. An examination of the shell of an individual of this group reveals that although descriptive names are sometimes misleading, in this case the name is fitting and apt. The box turtle has the ability to withdraw head, legs and tail entirely within the shell. The sections of the hinged plastron fit so neatly within the carapace as to "box" the turtle completely. The plastron is not united to the carapace by a bony union but is attached by ligaments and no solid intervening bridge separates its two hinged sections. The carapace, in all species of Terrepene, is blunt, oval, and more or less highly rounded. In general, there are no sculpturings or grooves on the shell other than the divisions between the shields.

Two groups of box turtles are found within the United States; the members of one group are possessed of four claws on each hind foot while those of the other have but three. The common eastern box turtle, *Terrepene carolina,* and the box turtle of the Middle West, *Terrepene ornata,* are in the first group.

In general appearance, character and habits, the several species are much alike, differing mostly in range and superficial coloring.

Sex differences, although not marked, are usually obvious to the knowing eye. In the females, the rear half of the plastron is smooth and flat, while in the males it is slightly convex. Males of the eastern box may also be distinguished by the bright orange-red iris of the eye, strikingly different from the dull brown of the female.

Although box turtles are classed as terrestrial, they do not shun water and are ordinarily found not far from streams, ponds or swamps. In dry weather, the eastern box is apt to forsake woods and bushy pasture land for the mud of wet and swampy places. The young, especially, like dampness and in captivity have a fondness for soaking in their water dish. The woodland terrarium with its mosses, ferns and cool hiding places, has proved in practice an ideal home for such youngsters. Adults kept under similar conditions would not be likely to survive. Although they like an occasional soaking, mature box turtles confined to small, damp quarters with insufficient sun, quickly succumb.

Although spotty in their distribution through a given locality, box turtles are to be met on a trek through the woods and fields. They have no scent that humans can detect, but apparently have an odor similar to that of one of our game birds which the sensitive nose of dogs can pick up. According to a veteran hunter and dog trainer, bird dogs often point a concealed box turtle and, as often, look most foolish and disconcerted when the object of the point is revealed.

In the matter of food the box turtle is not hard to please. In

the wild, a diet of insects and their larvae, slugs and snails,
earthworms, small fruits and berries and other vegetable material meet with its approval. Cantaloupe growers complain that
these turtles have the bad habit of ambling through a patch,
taking casual gouges out of melons and thus spoiling a considerable number. A box turtle that shows poor appetite in captivity often deigns to begin feeding if a bit of cantaloupe is included in the ration.

Mushrooms are a favorite with them and they are, curiously
enough, like the squirrels in that they are immune to the
ill effects of even such deadly fungi as the fly mushroom,
Amanita muscaria, and the destroying angel, *Amanita phal-
loides.*

THE TORTOISE

The most obvious distinction between the true tortoises and
land-dwelling turtles, such as the box turtles, is in the feet.
Although the box turtles are almost entirely terrestrial, they
can swim and their feet, with the slight webbing, still show
their close kinship to the water turtles and terrapin. The true
tortoises are almost helpless in the water and have been known
to drown. Their feet have no vestige of webbing, but are club
shaped and equipped with heavy, digging claws.

Somewhat unflatteringly, but with truth, the word tortoise
—derived from a Greek root meaning to twist—refers to the
bowed and clumsy front legs of this group.

The greater number of species of true tortoise are found in
the Old World. The United States has but three within its
borders: the desert tortoise of the Southwest, the gopher turtle
of Florida and the Gulf States, and Berlandier's tortoise, found
in southwestern Texas and northern Mexico. All are much
alike in appearance, living habits and disposition.

The carapace of the southwestern desert tortoise, *Gopherus
agassizii,* is fairly high and rather flattened on top. In color it
is dark brown or blackish with perhaps a hint of yellow in the
centers of the shields. The plastron is not hinged but is rigid
and firmly united to the carapace. The tortoises are unable to
attain a protective closure of the shell. When disturbed, the
head is drawn in with a hiss of expelled breath; the front legs
pull in after and the tough and scaly knees, coming together
in front, close the opening. The hind legs are pulled in so that
the rear opening is closed by the feet.

Robert Snedigar

GOPHER TORTOISE

Robert Snedigar

SOFT-SHELLED TURTLE

ALLIGATORS IN THE CHICAGO ZOOLOGICAL PARK

The gopher tortoise of the South, *Gopherus polyphemus,* is in every way similar to the desert tortoise except that the fore-legs are much more heavily scaled and have a ridge of spines down the front.

Both species are burrowers, as their heavy claws would indicate. The gopher tortoise of the South inhabits dry, barren and sandy wastes. Likewise, the desert tortoise is found in the most arid parts of the Southwest. This does not necessarily mean that the tortoises are unduly fond of sun and can stand an unlimited amount. Their habit is to stay within the burrows during the heat of midday and forage for their vegetable food during the early morning hours or when clouds cut off the most intense sun. One midsummer, after carrying a ten-inch gopher tortoise from the Okefinokee Swamp to Miami, we put the animal in a seemingly suitable enclosure for the night. Having much to do in the morning, it was late before any of our party bothered to look at the tortoise. The day was not hot, but the pen was poorly shaded and the sun, beating down upon concrete, had built up by ten o'clock enough heat to kill.

Both the desert tortoise and the gopher are almost strictly vegetarian, but will take some animal food if available. In captivity, a diet of canned dog food (rice or barley cooked with meat) with generous amounts of lettuce and available fruits is acceptable. Adult tortoises need dry, airy and clean quarters, with plenty of sun and plenty of shade. Young tortoises do poorly if kept in too dry a pen or cage. Slightly moistened sand is needed for them.

The burrows of the gopher tortoise offer refuge to other animals. The large gopher snake, Drymarchon, often evades capture by slipping into a tortoise burrow and the gopher frog, *Rana capito,* is seldom found anywhere else. The Florida variety of the white-footed mouse has been found nesting in burrows. Spiders frequent the entrances; various insects live within. As far as the tortoise is concerned, these other tenants are perfectly safe. Such dissension and devouring as go on in the burrow will be strictly between the guests.

THE SOFT-SHELLED TURTLES

One group of turtles is an exception to the general rule that turtles are possessed of a heavy body shell into which the soft parts may be more or less retracted for protection. These have shells, but, instead of being of bone with a covering of horny plates, they are soft and leathery. The North American genus of soft-shelled turtles is that of the Trionyx which, in four species, was formerly supposed to be distributed only throughout the Mississippi and Rio Grande Valleys and the southeastern states. From time to time, reports have been made of soft-shelled turtles also occurring in Pacific coast streams, notably the Colorado River. The Chinese have a liking for the flesh of these turtles (this liking is not exclusively Oriental—Trionyx are in demand in southern markets) and, lacking the evidence of actual specimens, it was assumed that perhaps one of the Chinese species had been transplanted to western regions. This may still be true, but two newly hatched specimens from the Colorado certainly belong to the species of the Rio Grande watershed.

These young are not much larger in diameter than a silver dollar and, of course, are harmless. Adults of this species grow to be very large, have a nasty disposition and an inclination to indulge it. The similar southern species, *Trionyx ferox,* sometimes as large as eighteen inches in diameter, is equally vicious. The relatively exposed underparts and the soft carapace—so soft that the outlines of the ribs show through—need the protection the knifelike jaws give them. These turtles, like other reptiles, never invite combat, but if unable to escape in flight make the aggressor realize, by their fierce battling, that he's caught something. Fishermen often hook them and, unless wary and lucky, are likely to need a few stitches in a finger—or even a new finger, if the captive is large. For the soft-shelled turtle's neck is long and snake-like; swift is the darting thrust of its head and the slashing snap of mandibles.

In habit, the soft-shelled turtles are almost entirely aquatic, but like and need their regular sun-bath. Hence, in addition to plenty of swimming room, they must have something in the way of a ledge to crawl out on. This should be of old, water-soaked wood or other soft and non-abrasive material. The plastron and the floppy edges of the carapace, once scratched, offer an entry to the fungus infection so common in turtles. Overcrowding, with its resultant clawing and fighting, should be avoided for the same reason.

One natural food of adult soft-shelled turtles is fish, so their feeding in captivity offers no great problem. Hatchlings and very young seem to prefer tubifex and white worms to chopped earthworms and to be able to recognize such living food a little more readily. Feed daily.

In summer, turtles of any size of all species are better off outdoors where they may receive plenty of sun. An enclosed pool with a sand beach area for sunning is ideal and, outside of labor cost, not expensive. If the matter of drainage permits, the pool should be below the surface of the surrounding terrain. The excavation may be of almost any shape the designer fancies or which fits into the general landscape. The outer wall, of rockwork or fieldstone, should be a few feet above the ground on the outside. On the inside of smooth cement-finished concrete, it runs down to form a ditch or moat all around the bottom of the excavation. An irregular mound left in the center of the hole provides an island. The concrete of the moat

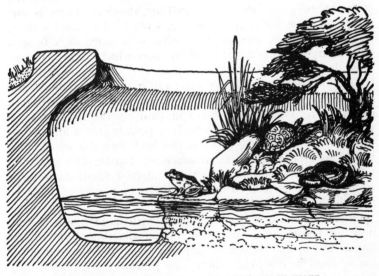

PIT CONTAINING BULLFROG, TURTLE AND BLACK SNAKE

should be carried up the sides of this well above the intended water line. This level will probably be about the same as the ground level outside. The island can be built up higher with earth taken from the moat excavation. If possible install a drain in the moat. A water connection is valuable. If running water is installed, it will be necessary to make some provision for an overflow pipe at the level of the island to keep the whole thing from flooding.

If large enough the island may be planted with small shrubs, clumps of wild rice, cattails and other aquatic plants grouped along the margin and unobtrusively protected from the turtles by wire mesh. Aquatic turtles will be happy in the moat while box turtles will find the island comfortable if provision has been made for shade during the heat of the day.

If the inner edge of the wall is carried well over so as to make a good overhang, the pit may be used for harmless snakes as well as turtles.

If such a pool is out of the question, round cypress wood tanks of the type made up for stock watering troughs, while not elegant in appearance, are serviceable and to be preferred to rough concrete.

Diseases of turtles are difficult to treat, for, as is true of most of our animals, by the time we recognize that something is wrong, it is too late to do anything about it. There is one exception. The shells of turtles, particularly the under surface, frequently become infected with what is presumably a fungus. The horny covering is eaten off in spots which, unless treated, grow larger and larger and expose the raw bone beneath. Rough concrete is one of the worst causes of this trouble, as the continued scratching breaks the shell and permits the entrance of infection. Treatment consists in painting the infected spots with a saturate solution of potassium permanganate and allowing it to dry before the turtle is put back into the water. This should be repeated as often as necessary. Turtles with adequate sun, in non-abrasive tanks, are rarely affected. Small turtles kept in aquariums need a walnut-sized lump of plaster of Paris to provide the needed calcium.

Occasionally box turtles are found with large lumps on the neck, or behind the legs. These lumps sometimes contain the larvae of a fly and should be carefully watched. When an opening appears in the lump, slit it and remove the larvae with a forceps. They are ready for pupation and, if put in damp sand in a covered dish and kept at between 70° and 80°F., will burrow in, eventually to emerge as flies. These are very rare, even in large museum collections, and would be a prize for any entomologist.

The Alligator

THE AMERICAN alligator, *Alligator mississippiensis,* is among the several reptiles most likely to come into an unprepared household. In the South, young alligators are one of the most popular tourist souvenirs and curio shops make them a featured item. These homely and scary little monsters very seldom find life kind to them. Dumped in a bathtub or in a tank, they refuse to feed and unless turned over by a bored owner to the nearest zoo, die in desperation.

As pets, alligators seem to me lacking in beauty of form and color and amusing behavior. However, as our principal representative of an important reptile group, the crocodilians, they are very interesting and deserve tank room.

When William Bartram visited the South in the late eighteenth century he found ". . . the alligators were in such incredible numbers, and so close together from shore to shore, that it would have been easy to have walked across their heads, had the animals been harmless." In those days, old 'gators were reported as much as twenty feet in length. Today, even in the fastness of the Okefinokee and the Everglades, alligators are increasingly scarce and specimens of more than twelve feet in length decidedly rare.

In the protected areas of the Okefinokee, it is likely the alligator will be able to escape threatened extermination. In such swamps, it is thoroughly at home and well suited in temperament and appearance to the environment.

The rough, scaly back protects it from attack and, as the creature floats under the surface of the water or lies on the grassy bank, makes it seem just another bit of debris. The tail, flattened for swimming, is a powerful weapon and out of the water can be used with lightning speed to slap an enemy.

The natural refuge of the alligator is the water and in its depths he can sulk for hours without suffering. The home is in a deep hole dug in the marshy bank of slough or swamp. This reptile is quite sensitive to cold and commonly spends intemperate weather in its cave or dug into the mud.

The alligator's feeding is almost entirely in the water. To enable it to hold struggling prey without having its own lungs fill with water, two tight curtains of skin across the back of the crocodilian mouth effectually prevent water from flooding the windpipe.

While seldom noisy in captivity, alligators have a voice and the thundering bellow of a large bull 'gator can be heard in the swamps for some distance. Youngsters squeal and grunt, especially at mealtime.

In nature they feed upon all kinds of small aquatic foods and fish; in captivity their daily fare is meat, fish or liver. Young alligators quickly become tame enough to feed from the keeper's hand but their quick snap makes it advisable to tender the morsel with a stick or forceps. The creature's first impulse upon grabbing food is to tear off a chunk by a rapid twisting of the body. This does a finger no good.

The prime requisites of small alligators in captivity are sunshine and warmth. Without these, the best of food will be rejected. Naturally, the pen or tank depends upon the size of the inmate. Very small specimens can be satisfactorily taken care of in an ordinary aquarium with a sloping sand bench rising out of shallow water.

Larger 'gators may be kept in tanks in which water can be frequently changed. If not in sunshine for at least a part of the day, they should receive ultra-violet light treatments regularly. Concrete tanks, unless well surfaced with smooth cement, are abrasive and unfit for use. When water is changed, try to replace with water of the same temperature or serious upsets may occur. In summer, larger specimens may be kept in outdoor pools with a platform or sand beach for sun-bathing.

Special Foods

FROM TIME to time mention has been made in these pages of special and unusual food. Flies, meal worms, wax worms, tubifex, white worms, crickets and the like have been recommended. The mere mention of these foods is not enough.

Unfortunately, in running this sort of a boarding-house, markets are of indifferent value and the boarding-house keeper finds himself forced to raise or catch his own ham and eggs if he is to suit the needs of his guests. When flies and other insects are the ham and eggs desired, the problem is not always easy. During the summer months, live food can be caught by sweeping the grass of meadows with an insect net. This source of supply, unfortunately, is not available during the winter months when such food is most needed.

THE MEAL WORM

Of small live foods which can be raised, the most generally useful and easiest to keep in stock is the meal worm, the larvae of a black beetle, *Tenebrio molitor*. A start of these is obtainable at any good pet shop. The method used in rearing them is largely dependent upon the yield desired. If the demand is small, a large biscuit tin or other metal container is excellent. Where quantity production is not desired, keep one box which contains beetles, pupae, eggs and all sizes of worms. The can may be open at the top if the sides are slick but it is better to cover it over with screen. The food of the worms is bran and the can should always contain plenty of it. Two or three inches of it should be put in the bottom of the container and worms or pupae added. Cover the whole with two or three layers of moistened burlap to provide a holder for subsequent sprinklings. Burlap also considerably simplifies the matter of collecting worms for feeding. They tend to gather between the layers and may easily be scooped up in the hand or picked out with forceps. A favored place for the larvae to pupate is between the burlap and consequently picking them is simplified to some extent. As the worms go into the easily recognized pupal stage, watch for the appearance of beetles and when they begin to emerge, put in slices of raw apple, potato or carrot for their feeding. In a mixed box, bran and vegetable must be added from time to time and the burlap moistened once or twice a week. This arrangement satisfies small needs only.

A great drawback to a mixed culture of this sort is that it contains everything from eggs to beetles. The powdery waste from the bran eaten by the worms accumulates and sifting is necessary to get rid of it. Unfortunately, in this case, sifting means that eggs and small worms go through the fine sieve and into the trash can.

For controlled quantity production, the best practice is to remove pupae from the general culture as fast as they form and transfer them in lots of a hundred or more to prepared containers. There the adult beetles emerge at about the same time, mate, deposit eggs and die. During their short life they need food, and apple or potato should be continually supplied. To stimulate egg-laying and insure a lusty colony of worms, moistened ground dog biscuit or canned dog food may be given sparingly. Moisture is important during breeding and the burlap should be sprinkled regularly with a fine spray. Too much water in one spot cakes the bran and encourages the development of molds.

In due time, the bran will be found alive with small worms. As these grow and feed, new bran must be added. By the time the waste has become excessive, the worms are large enough to stay in the sieve and the box can be cleaned by sifting without loss.

As pupae form, new containers may be started until the colony is large enough to supply the demand. In the department of Experimental Biology of the American Museum, the meal worm colony consists of more than twenty wood-jacketed, galvanized metal trays, each approximately five feet long by two wide by one high. These are stacked in racks which permit a free circulation of air. This colony yields daily a full quart or more of worms of all desired sizes.

High temperatures (more than 80°F.) are not necessary but do speed up the life cycle of the worms and result in increased production. The life cycle from egg to beetle takes about four or five months. With proper care, feeding and watering, there need be no slack season after a colony is established.

Meal worms are generally serviceable as food, but the horny

outer covering affects the intestinal tract of reptiles if the worms are used to the exclusion of other live foods.

Meal worms do not live long in terraria or dishes filled with moss or other wet material, and unless they are picked out when not consumed by the inmate, foul the dish and furnish a start for mold. The meal worm *Blapstinus moestus,* smaller and more delicate than Tenebrio, has a greater resistance to moisture and has been recommended for amphibian feeding. Reared in similar fashion in similar containers, these worms are reported to grow fast and quickly produce a large colony. Moisture is important in rearing them, and to supply it, additional layers of burlap should be laid over the bran and a daily sprinkling given.

THE WAX WORM

The wax moth, *Galleria mellonella,* is a little more difficult to manage than the meal worm but is very useful, particularly if delicate lizards are to be kept in good health. Its larva, the wax worm, has no horny outer coat but is covered with a soft white skin. Continued feeding of it is not irritating to the intestinal tract.

The adult wax moth never eats. Its only function and desire in life is to mate and leave a goodly mass of eggs in some beehive. These eggs are laid in any part of the unfilled wax, but preferably in the old brood cells. The eggs hatch in about ten days and the tiny larvae eat their way into the bottom of the comb. As they progress, the comb begins to break down and piles of the characteristic excrement, like brown sawdust in appearance, are evident everywhere. The worms grow fast and immediately begin to spin aimless strands of fine silk. They have a tendency to bunch and a perceptible rise of temperature in the comb is apparent. Eventually, they turn their spinning to a practical purpose and make a cocoon in which to pupate.

In rearing wax worms, the first requisite is a supply of old brood comb. This is not always easy to obtain, but a check of apiaries should reveal a steady source of supply. Clean unused comb or comb in which honey has been stored *will not do.* The worms disdain it. Comb of a heavy texture and dark brown color, without honey or masses of stored pollen, is what is needed.

For a start, it is usually only necessary to get such a comb and keep it in a warm place for a few weeks. In most cases, it is already infested with larvae or eggs, as the wax moth is one of the bee-keeper's ever present problems.

Wooden boxes, twelve inches by twelve by twenty-four, with a tight hinged top, have been found to be satisfactory rearing chambers. Four-inch holes in each end, screened on the inside by fine copper cloth, are necessary for ventilation. Metal-lined boxes are not satisfactory. The larvae like to chew into the wood for a place to pupate.

For small needs, one box is sufficient, but if regular day-by-day production is desired, a series is necessary. Regular staggering, as with meal worms, produces the best results. Pupae, hand-picked from the first box (this is a somewhat tedious job), are transferred in quantities of a hundred or more to a second box well supplied with comb. This second box is set away in a warm place for a month or more before opening. By that time, most of the moths are dead—their life work done—and the wax is alive with small worms needing a fresh supply of comb. These maintain a fairly constant rate of growth and the culturist with a good series of boxes can usually get exactly the size of worm he wants up to the maximum of an inch. Spent moths are not useless but, by their movements, often stimulate frogs and lizards of captious and uncertain appetite into feeding.

Wax worm boxes need a good deal of routine care and should be gone over at least once a week while the larvae are growing. After the worms attain a fair size they pupate and unless taken out, upset any attempt at systematic production.

THE BLUEBOTTLE FLY

The ordinary bluebottle fly or blow fly, although unpleasant to raise and keep, is almost essential if large groups of small lizards and amphibians are to be kept in good health over periods of time. Their larvae or maggots have been used in hospitals in the treatment of certain bone infections and sterile maggots were formerly regularly produced by biological supply houses for sale.

Breeding for food does not require the precautions and clean conditions that medical use does. If flies are needed in enormous quantities every day, the following method of production has been found satisfactory and economical. For smaller needs, some modification would no doubt be necessary.

Breeding cages are rectangular with screen sides, lift-up

screen top and wooden ends. In one end, several holes are bored to receive vials for egg-laying. A batch of a hundred or more adult flies or pupae are put into the cage to lay. Their food is a half-and-half mixture of beaten egg and slightly sweetened water. This is put in at least once a day through one of the vial holes by means of a suction bulb and glass tube. Crumpled paper towels in a dish keep it from the floor of the cage. Small cubes of beef heart are impaled on a toothpick, placed in a vial and the vial put in one of the openings. The flies, when ready to lay their eggs, will quickly avail themselves of it.

When two or three good-sized egg masses are in a vial, remove and replace with another. The eggs are taken out of the bottle and placed carefully on a thick slice of fresh beef heart two or three inches in diameter. This, in a glass dish, is placed in the center of a sawdust-filled ten-inch wide, straight-sided glass vessel. The eggs hatch within twenty-four hours. The growing larvae will need more meat before they reach adult size and it should be supplied as needed. When fully developed, the maggots leave the meat dish and pupate in the sawdust. The pupae are separated out by sifting, and placed in small gauze-topped bottles or jars in appropriate numbers. In three or four days the fully developed flies begin to emerge. Gauze held by rubber bands is not only a good covering for the pupae bottles but is useful to keep maggots from leaving the sawdust dishes, as they occasionally do if moisture conditions are not quite right.

Temperature is very important in fly-breeding. Best results are obtained at between 70° and 80°F. Above or below that range, the eggs are sterile and will not hatch.

In England, fly maggots are called gentles and are raised for fish bait. In most cases, our animals prefer the adult fly, but for some lizards and frogs the maggots themselves may be used. In such cases, they should be washed and put into damp sand to scour for a few hours before use.

During the warmer months flies can be caught in the old-fashioned wire funnel trap with excellent results. Bait with slightly used fish or meat.

FRUIT FLIES

For very small lizards and amphibians, the fruit fly, *Drosophila melanogaster,* furnishes an infinitesimal but tasty morsel. Many techniques of Drosophila-rearing have been developed.

These flies are used extensively in laboratories for working out and demonstrating genetic problems in which a large number of generations of controlled heredity are essential. Many forms have been developed, among them a wingless type. This, if available, is particularly useful for feeding as it is easier to handle and does not escape from cages as readily as the winged types.

For experimental use, Drosophila-raising is carried on under stringent conditions. Vessels and foods are sterilized and great care is taken to avoid any possible contamination. In raising them for food only, less care is necessary and relatively simple methods and ordinary temperatures suffice.

If cultured flies are not available, a start may be obtained in spring and summer by exposing a quart milk bottle containing a few pieces of well-ripened banana. Even in the city flies will be attracted by it and deposit eggs. From this beginning more systematic production can be started.

Crumple filter paper or a paper towel and place in a clean milk bottle. Put in half a banana cut into quarters and a dozen or so adult flies. Quickly close the bottle with a wad of gauze-covered absorbent cotton. The flies breed quickly and in less than two weeks the bottle will contain a colony ready for use as food. The simplest method of feeding is to put the bottle in the lizard or frog cage and pull out the stopper. The cage, of course, must be fly-tight or covered temporarily with a glass. In transferring flies for breeding from bottle to bottle, take out the gauze cork, put the bottle mouths together and tap the one containing flies. When a few have passed over, recork both containers.

THE EARTHWORM

Among the most generally useful foods for small snakes, frogs, toads, salamanders and shrews is the common earthworm, *Lumbricus terrestris*. During the warmer months of the year, earthworm collecting is not a difficult job. Rainy nights bring the large night crawlers to the surface of the ground and often a couple of hours' work with a flashlight will provide a month's feeding. The worms are not especially sensitive to the collector's light, but are very touchy about ground vibrations and too heavy a tread spoils the game. The best collecting ground is a well-kept lawn. If the individual has no liking for paddling around in a warm rain, digging worms in the time-honored fish-bait fashion is the only alternative.

Hand-to-mouth feeding is all very well for summer but, if earthworm feeding is to be continued over winter, steps must be taken in the fall to lay in a supply. Large boxes—stout enough to withstand the effects of moisture—or, still better, half-barrels, filled with rich sandy loam and set in a cool place —are good containers. Soil with a clay content is to be avoided. The best mixture is made from equal parts of sandy loam, rotted wood and well-rotted leaf mold. Before putting in worms, the soil mixture should be thoroughly dampened and tossed with a shovel several times.

As worms are collected, dump them into a depression on the surface and cover over with several layers of dampened burlap. Healthy worms will dig down into the mass. Those injured will remain on the surface and may be removed later. Moisture must be watched. If kept too wet, the worms come to the surface and crawl out. If too dry, they perish.

If worms are to be kept for short periods only, feeding is not necessary. There is enough nutriment in the leaf mold to satisfy temporary needs. If long storage and breeding are contemplated, feed weekly with ground raw potato or apple, uncooked oatmeal moistened with milk, stale bread and milk, cooked potato or other starchy material. The quantity must be suited to the size of the container and the number of worms estimated to be in it, but should not be excessive. Unconsumed food indicates too heavy feeding. All food should be broken up and lightly buried beneath the surface of the soil. Left on top or buried in lumps it starts molds.

If conditions and temperature—about 70° F. is breeding temperature—are right and a goodly number of large worms is in the colony, tiny earthworms will be found under the burlap within a few weeks. These tender morsels are greatly relished by small tree toads and tropical fish.

The superficially similar manure worm is often mistaken for the earthworm and collected in its place. Living largely in wet cow manure, these take from their habitat some rank and offensive quality which renders them practically worthless for feeding. A few salamanders and small snakes will take them for want of anything better but creatures of delicate palate reject them.

WHITE WORMS

For the smaller amphibians and fish, the white worm, *Enchytraeus albidus,* is invaluable. Tropical fish dealers or pet shops can usually supply a start for a white worm culture. The raising of white worms, while not a highly technical matter, is one in which care and consistent attention are important. Very often, through sloppy handling and neglect, a flourishing colony dwindles and dies. Too much water or too little; heavy and careless feeding with a resultant growth of mold; lack of aeration; over-heating; all may contribute to the disaster.

We have found the best container for white worms to be a wooden box approximately fourteen inches by ten by ten. The culture medium is a light garden loam, without too much sand, mixed with equal parts of rich, well-rotted leaf mold. If possible, sterilize the mixture by baking in an oven. After sterilizing, add water carefully. The degree of dampness is important. When a handful squeezed lightly holds together, the water content is about right. Beware of too much water; the earth should not be sticky. In order to mix in the water thoroughly, it is best to put the whole mass on a board and turn over and over with a small shovel in the manner of a man mixing cement.

The start of white worms may be placed in one box or if large, split among several. First fill the box half full of the prepared soil. Scatter the worms over the surface and cover over with a couple of inches more of dirt.

Food for white worms may be varied. Stale bread, moistened with milk but not sloppy; raw oatmeal and milk or other cereals are satisfactory. An alternation of foods is good practice. Feeding should be done weekly and at the same time the culture aerated and moistened. This is best done by turning the whole box into a large dish pan and tossing the mass like a cook making biscuit. This breaks up lumps. Worms will usually be found gathered in bunches and at this time may be picked out easily. Put in a jar with a little earth and covered with gauze, they may be kept in the refrigerator for the week's feeding and the culture need not be disturbed again.

When the earth is well stirred up moisten, if necessary, with a fine bulb spray and re-mix. Pour two-thirds of the earth back into the box and sprinkle the food—beware of too much—over the surface. Cover with the rest of the earth. A damp piece of muslin laid on the box prevents drying. If kept in a dry place, it might be well to cover the box over with a piece of glass in addition.

The white worm colony should be kept in a cool and dark place—60°F. or lesss—for best results.

Worms for feeding are easily cleaned of earth by two or three washings in cold water. Their tendency to bunch makes it possible to lift them out in clean masses with forceps.

TUBIFEX

Shallow fresh or brackish waters, well bottomed with muck and mud, form the breeding grounds of the small red worms of the family Tubificidae. They may be found growing in great numbers in almost any place where sufficient amounts of organic matter to support them are deposited.

Large tropical fish stores have tubifex in stock at all times. Their supply is replenished regularly by professional collectors. These men, gathering large quantities, necessarily choose rather unsavory localities in which to do their work.

The hours just before and after dawn and dark are the best collecting times. When feeding, tubifex stand head in the muck with tails aloft and waving in the water. If they are not seen on the surface, a little washing of suspected muck shows whether the particular site is worth bothering with or not.

The upper layer of mud is scooped up with a dip net, put in a sieve and superficially washed by pouring water through. Sticks and coarse debris may be removed at this time. The residue consists of about four parts mud to one part worms, and must undergo careful cleansing before use. The mass should be washed in a pan with running water—the worms tend to bunch and sink—with a rotary, sloshing movement. The final step in cleaning is to spread the remaining muck and worms on the bottom of a deep can and cover over with a couple of inches of clean sand. Fill the can with water from a hose, being careful not to bore through the sand layer, and let a thin trickle run into the can for some hours. In time, the tubifex will have left the mud, wriggled up through the sand layer, and lie on it, a solid red mat in continuous rippling motion.

Because of the large quantities of decaying organic matter necessary to maintain them, the breeding of tubifex is practically out of the question. They may be stored for some time by burying a cleansed mass in wet sand and keeping in a cold place. Kept in cold running water, they last for a couple of weeks. In this case, the worm mass should be broken up with the hand at least once daily to allow dead worms to float off.

Tubifex are the mainstay of the tropical fish fancier who

needs quantities of live food. Small amphibians and baby turtles thrive on them.

THE CRICKET

The house cricket, *Gryllus domesticus,* in addition to being
a valuable food insect, is rather nice to have around for its own
sake. The chirping little noise is not unpleasant and provides a
genial undertone to general activities. In the Orient this song
is appreciated and caged crickets—the cages often elaborately
wrought and decorated—are a part of most households.

The cricket-raising tank or jar should be fairly large and
provided with a screen top. A battery jar, twelve inches wide
by fourteen or more high, is excellent. The bottom should be
covered with at least two inches of sterilized sandy loam over
a shallow drainage layer of gravel.

In starting a colony, six or eight males and as many females
are needed. Females can readily be identified by their long
ovipositors. The eggs are buried in the earth and when young
appear, the adults should be removed to a new container.

Bran mixed with powdered milk is the principal food
and a small dish of it should be available at all times. In
addition, scatter small pieces of carrot, apple, potato and lettuce over the jar bottom. A little puppy biscuit or dried meat
furnishes protein, and prevents cannibalism. A part of a cake
of old-fashioned dry yeast—not compressed yeast—kept in a
dish by itself, stimulates breeding.

Mixed batches of different sizes should not be kept together.
As with other live foods, a staggered series gives the best results.

Daily light sprinkling is necessary but care should be taken
not to saturate the yeast and food.

ROACHES

Roaches are not particularly pleasant customers to handle
and work with, but they offer a relatively easy source of food
for amphibians and reptiles. A start of roaches is not the most
difficult thing in the world to procure, especially in eastern
cities. Two species, the large American wood roach and the
smaller roach sometimes called the Croton bug, are often pes-
tiferously common.

Given a chance, the latter multiplies rapidly and quickly
overruns the infested premises, to the horror of housekeepers.
The large roach is more likely to be found in basements, ware-
houses and similar dark and damp locations.

Either species may be raised or trapped for use as food. Trapping is simple. A gallon can or straight-sided jar is baited with bacon rind, banana peel or other fragrant garbage. A two-inch band below the rim on the inside of the jar is smeared lightly with vaseline. Set in a place frequented by roaches, such a trap yields a surprisingly large catch. They fall past the grease in their eagerness to get at the food, but are unwilling or unable to crawl over it to get out again.

Although raising roaches involves a little time and trouble, it is the only safe practice if roaches are to be used as food. Trapped roaches have been feeding on all sorts of unknown material and may even carry on their bodies poisons set out for their destruction. The container for raising roaches is similar to that used for crickets. The upper part of the inside of the jar is smeared with vaseline to prevent escapes when the screen top is taken off. Several layers of crumpled blotting paper in the bottom furnish hiding places. The food may be varied according to what is available. Feed about three times a week: potatoes, carrots, scraps of lettuce, bread, apple, bacon rind, etc. Old-fashioned dry yeast should be kept in the jar as for crickets. Spray lightly once a day.

Roach cages should not be kept in too light a place, but preferably half dark. Warmth promotes heavy breeding.

Twenty-three Years After

THE MAJOR interest of early naturalists was in the classification of plants and animals and the placement of each species in its proper niche. This is an essential activity, certainly, but for years it so occupied the minds and time of most students of plant and animal life as to overshadow almost completely many other interesting fields of investigation.

The study of animal behavior was an inevitable part of the life of primitive man and a knowledge of the habits and ways of wild creatures and the properties of plants was essential to survival. The uncanny awareness of the Australian aborigine of the effects of subtle changes in conditions on the game on which his life depends, has often been noted. However, in our own culture, our knowledge of the wild has been of less immediate importance and somewhat neglected. A great deal of our wildlife lore has been just that—folklore loaded with tall tales, misapprehensions and even nice, fat lies. With the development of psychology and the scientific study of the human mind and brain, men, here and abroad, began to devote time to ordered studies of animal behavior with rewarding results.

From the works of relatively recent years, there have come ideas that the animal keeper can use for the mutual benefit of himself and his charges. The study of the interlocking needs of the creatures of a particular environment and their effect on the environment itself has long been merely a fuzzy and not well understood mass of information. Utilization of new findings in geology, botany, chemistry, zoology, and many other sciences has brought into being and recognition the science of ecology. Ecology deals with life and the manner of its living and no facts pertaining to life and the shifting influences that affect it are foreign to the field. The relationship between atomic fallout and embryology is only one of the very important ecological studies necessary today.

Among zoologists, precise and systematic surveys and studies of animal populations and individuals confirmed some of the inherited ideas about animals and disproved many of the more fanciful notions. The "nature fakers" of the early part of the century began to have a tough time when confronted with exact data and have almost vanished from the scene. A few sturdy traders in "abominable snowmen," etc., pop up once in a while but even a gullible public now demands proof of wonders.

Very early observers had noted that many animals had definite travel patterns and as individuals, pairs, or clans occupied certain areas to the exclusion of others of their kind. The expression of a somewhat exaggerated feeling for property rights has been made evident to most of us at some time or other. The social insects—bees, ants, wasps—while not establishing a territory to defend food-providing areas against their own species, have a powerful defense against invaders of the nest or its immediate vicinity. Many mammals, birds, fish, and reptiles establish strictly defined territories during the breeding season. This area is defended by the resident male against other males and, very often, onslaughts on possible predators can be sudden and, especially when the invader is a human, most disconcerting. There have been cases publicized of robins, nesting under eaves, which have successfully intimidated owners of property and denied them access to garages, summer houses, and other facilities. Defense of the nest and its environs is not to be confused with the defense by the male robin of the larger territory which constitutes the feeding range. The robin and the singing birds generally use their song as a territory marker. The male cardinal, high in a tree, is not bursting his throat to express great joy and nuptial bliss but is proclaiming to all cardinals in range of his voice that the territory is taken and he proposes to defend it if necessary. Male birds, cardinals and robins especially, often engage in vigorous duels with their own reflections in shiny hubcaps, windows, and other reflecting surfaces. The rivalry is, of course, not only over territory but also has in it a good dash of sexual jealousy. The territories of nesting birds ordinarily become less important to them as the young hatch and mature. However, there are many cases of birds protecting winter feeding areas.

Birds are not the only creatures to define the limits of territory by sound. The howler monkeys of the American tropics and the lar gibbons of Asia live in clan groups, sometimes of fair size, that proclaim by concerted and far-carrying vocalization of a whole group that the area is already occupied and not open to homesteaders.

While the rolling bass roars of the American alligator have a territorial significance, in these thundering crocodilian cacophonies there is another message: Young gentleman alligator, established comfortably in own puddle, would like to meet lady alligator; object: matrimony. Reptiles do not establish territories as nesting pairs. Males take over a desirable area and allow females to enter it freely but vigorously defend their rights against other males. It is interesting to note that in some lizards and fish, there is good evidence for the belief that the interloping male in an occupied territory suffers from an inferiority feeling which cramps his style and lets him be routed and expelled by a smaller male.

The birds are able to make their presence known to all comers by the male's bright plumage and song; lizards have flashes of color and threatening behavior and appearance; but many of the mammals use still another means of marking out boundaries—scent. Droppings in strategic locations or secretions from special glands rubbed on tree trunks and other surfaces, while quite unapparent to the human passerby, are pungent signals to other animals—not only of their own species—of a presence and a claim.

Many animals with established ranges—the immediate vicinity of the den or nest and the surrounding food supply territory—in addition to fighting off attempts of outsiders to muscle in, have the problem of evicting from the family bed and board their own grownup offspring. A coyote youngster, after having received the best education his parents could provide, finds himself, along with his brother and sister littermates, cast out and on his own. His training has been good and his intelligence is considerable and, although he may have to travel some distance and undergo many hazards before he finds an area not overpopulated with his kind, he stands a very good chance of making a fine job of living.

Within any family or group of animals, certain relationships inevitably develop involving the personal rights, privileges, and restrictions of the individual members. A hypothetical human family might be a fair sample. Father, the Boss, we would call the "dominant male." Sometimes he is. Mother ideally takes precedence after him, and the various children in a series of steps, not necessarily determined by sex or age, have each a place in the pattern. Naturally, these relationships are subject to constant challenge and change. A challenge may come unexpectedly to a higher-up from the most humble member of the household or group. I am not too sure if the case fits exactly into our situation or not—as there were undoubtedly unknown factors involved—but one family dog, adored by the mother and devoted to the children, had been in the habit of giving up the comfortable living-room chair it fancied as soon as the step of the Boss was heard in the hall. One day it decided to stay in the chair and defied dispossession by Poppa with snarls and a show of teeth. The revolt, squashed by Mother, was a failure, but there still exists some competition between Boss and dog.

The type of animal group dominated by a male with a descending hierarchy of females, other males, and young was studied many years ago in Europe and dubbed a "pecking order" because the investigator chose as a convenient object for study a hen yard. The dominant male, the rooster, has first place and the right to chastise by pecking any other member of the flock. So much for his lordship. A hen, not necessarily the largest or strongest, may peck any other hen in the yard. But she may not peck the rooster. We might refer to her as Hen A. Hen B may not peck Hen A, but she may go down the line and deliver punishment and take precedence over other occupants of the yard. Hen C may not peck A or B but may peck D, E, F, and so on down the order. However, the chain is not always so nicely established. A hen low in the scale, Hen S perhaps, may be dominant over B or C while still remaining subservient to A, D, E, and down to and including R. Down at the bottom of the order, it is not uncommon to find a poor fearful creature, pecked by everyone and afraid to peck anyone.

In an animal family, clan, or herd, in addition to social status, high or low, good or bad, the individual may have personal rights of a sort such as a favored resting or roosting place. Clans often have within them small companies made up of individuals who like one another's company. In baboons, these may be a few young males traveling more or less on the outskirts of the major clan and subject to the constant suspicion and scrutiny of the dominant male. In a baboon clan, a mother with a baby is usually in the company of several females without offspring who share in the protection, grooming, transport, and care of the youngster. Within such a group, dissensions often arise as suddenly as a summer storm, accompanied by screams and squalls of rage and expostulation. It has been noted in the wild that when such a tumult breaks the quiet of the group's travel and feeding, the dominant male, although he may be some distance away, will rush to the scene to find out what the fuss is and to restore order. Many clan animals have great willingness to unite to fight off enemies, and perhaps baboons have this feeling of solidarity more than most. No one, human or animal, can quarrel with a single baboon. The fight is with every baboon in sight or hearing.

In addition to group territories, many animals have individual places of relative safety which they frequent for resting or sleeping and seem to regard as peculiarly their own. In captivity, this trait provides the intelligent wild animal keeper with an instrument he can use for both his own and the animal's advantage. A lack of recognition of the animal's feeling that it has a right to a small place of its own has in many cases resulted in disaster for the careless keeper or visitor. Larger carnivores, especially, come to regard a cage or a particular place in a cage as personal property not to be trespassed on. In the case of the larger animals, any invasion of an animal's cage by a person normally accepted as a friend or as dominant over the animal is very hazardous. A man of my acquaintance—I didn't know him so well that I had to send flowers—a bear trainer, in a hurry to prepare his performers for their act, instead of turning a big bear into a neutral cage to put on his decorations and muzzle, brashly entered the animal's cage and aroused it from sleep. The result—a coroner's inquest.

Good trainers who work their performers without brutality or weapons respect the animals and, often intuitively, avail themselves of seemingly unimportant facets of the animal's nature to handle them successfully. Many animals, dogs and baboons for instance, with good peripheral vision may be obviously uncomfortable when stared at. An old keeper friend of mine, notably capable with many kinds of animals, said that one of the best ways to get along with critters was not to try to look them straight in the eye. A straight stare can be regarded as a threat or an invitation to a fight, when directed by one dog at another or one baboon at another. So, if a human persists in making threats, he may have trouble.

The good trainer or keeper is dominant over his charges. In an animal act, even when put on by a poor trainer, when the act begins, the trainer is in the arena before any animals are allowed to enter. He is there first. He is dominant. The cage is his and the animals, as intruders, are at a psychological disadvantage.

Another tool the intelligent animal keeper can use to advantage is an appreciation of "flight distance." The flight distance of an animal may be roughly described as the shortest space between itself and a possible enemy that the animal can permit without recourse to flight or combat. Flight distances vary greatly with different species of animals depending on natural speed, protective devices, food habits and, of course, temperament. Many creatures become extinct within historic times because, in an environment that was isolated and free of predators, they did not know enough to get up and go or fight back when Man, the arch-killer, appeared.

Animals of great size and physical power or with specialized protective devices, obviously will permit a closer approach than creatures devoid of strength or means of offense. The skunk, obviously, needs only a short flight distance while foxes and coyotes, with no special protective means, have to be wary of any other creature's close approach. In the wild, deer and antelope are off at the slightest hint of danger. Survival of many creatures is so dependent on flight that at first thought it might seem impossible to ever overcome the ingrained flight impulse and condition them to a captive life. But—life is full of happy "but's"—there enters into this matter of adjustment to captivity a factor of great value. Delicate and timid creatures can be successfully kept in captivity for long periods of time because, after the critical period of capture shock, replacements for the flight distance become acceptable. In the new conditions, fences, dry or water-filled moats, bars, screens, and glass take the place of the normal flight distance. The adjustment is gradual but most captive animals eventually seem to regard their barriers as existing for their protection rather than for the protection of the rubbernecking *Homo sapiens*. With larger captive animals, a failure by humans to appreciate the fact that a substitution has been made can cause disaster. More than once, a zoo visitor, desiring more intimate contact and assuming an animal to be perfectly docile and tame because it deigns to allow petting through the bars or over the fence and even accepts food from his hand, hops the fence or opens a door to a cage or enclosure and finds himself assaulted by a wild and raging beast. In one act, he has violated two animal needs. A territory has been invaded and an animal's flight distance cut to nothing.

The importance of a satisfactory substitute for actual flight distance as well as the skilled exercise of understanding of animals may be seen in the story of a mixed animal act of a few years back. The trainer, a most proficient man who scorned the use of guns and whips, wanted to place on adjoining pedestals a lion and a bear. The bear gave no trouble but the lion, not trusting his neighbor, balked and refused to take his proper place. The arena was of the usual circus type, vertical steel bars forming a circle. The trainer solved the problem by attaching to a vertical between the lion and the bear a section of birdcage wire which jutted into the arena a couple of feet. Though not visible to the audience, this fragile and seemingly inadequate barrier between the two pedestals was acceptable to the lion, and the act went on.

These matters of flight distance and territory may seem to have little to do with the care of small creatures, but any captive animal, large or small, timid or bold, in addition to proper food and housing, must be able to feel safe and secure and in a place of its own. Continual disturbance, limited space, and a lack of feeling of security result in losses of valuable and interesting animals in zoos and private collections every year. In a limited space, cramped and confined, with traffic all the way around, the most lethargic of beasts is apt to go "stir-crazy" and perish. With one or more solid walls to back up against, the situation is improved and, if adequate space and some area of seclusion is provided, even very delicate specimens adjust well.

One of the great rewards of the animal owner is the successful breeding and rearing of young. Only well-kept wild stock in good condition reproduce in captivity so that the birth of young is not only a reward but an award for excellence in zoo keeping. A chance to watch and even participate in the rearing of creatures of little known habits can be a very satisfying experience. Most fish, reptiles, and amphibians discontinue interest in their offspring as soon as the eggs are laid, but in the case of the mammals and birds, the birth of young or the hatching of eggs is only the beginning of an arduous task of feeding, cleaning, and education.

In this early stage of life, a young animal or bird is open to influences not acceptable in later life. Most infants will take their early associates to be of their own kind and will regard as parents whoever cares for and feeds them. This has been called "imprinting"; attachments formed by such infants are often difficult to break in later life. Dr. Lorenz tells the story of a peacock chick, the sole survivor of a storm-killed clutch of eggs, that was placed in the warmest house of a wartime European zoo. The immediate associates of the chick were giant tortoises, and with them he grew up. He ate with them, slept with them, and they were his kind. When mature and placed with other peacocks, the tortoise-imprinted bird wanted nothing to do with them and continually sought the company of his early companions.

Animals raised from a very early age by humans often have a rough time establishing relationships with their own kind even when the re-introduction is carried out gradually and diplomatically. The story of Elsa, the lioness, and the troubles of her human foster parents in returning her to a natural lion life is an example. Elsa wound up by living in two worlds, and although she passed from one to the other and back again, the two did not really overlap. Her cubs, although brought within view of her beloved friends, were not allowed to come to them. Similarly, I have seen a mountain lion, once a lovely pet, a she-wolf that had been hand-raised and was still most responsive and affectionate, and a coyote, very tame and full of fun, all keep their growing young to the rear of their cages and away from contact with human beings. And the young grew up as wild as or, in the case of the mountain lion cubs, even wilder than if their mothers had never had close human associations.

Imprinting can have great advantages in particular situations. Pups destined for service in the leading of the blind were found much easier to condition and work with if their early life had been with human foster parents rather than in normal litters of brothers and sisters nursed and educated by their natural mothers. I understand that the services of 4H Club boys and girls have now been enlisted in the hand-rearing of dogs for this need.

Work with infant animals has given the medical world substantial help in the treatment of not only physical, but also mental ills. Baby monkeys, reared with adequate warmth and food, but deprived of physical contact with a mother or substitute mother, have been found to be severely retarded in many ways and, very often, completely unable to adjust to any sort of normal life. Babies need more than sanitary surroundings and sterile bottles to grow up happily.

Many young animals need handling and affection. Others do not and may even resent having attentions forced upon them. Each individual has his own needs to be filled and all we can do is try and find out what these are and fill them the best way we can. Kindness, proper food and care, space and security and a recognition of any special requirements will usually maintain any animal, large or small, common or rare, in a satisfactory way.

In Conclusion

THE ANIMALS discussed in these pages by no means comprise the list of native creatures the amateur can maintain successfully in a happy captivity. The exigencies of space and the restrictions imposed upon the writer by his own experience and preferences have inevitably influenced the selections. This book can never be complete but it can have a continuation. *Your* notebooks, *your* photographs and *your* fun will be, I hope, that continuation.

There is just one more thing for me to say before I leave you to your job. It is simply that we must steadily keep in mind that animals are reasonable and that, within the limits of their philosophy and knowledge, they behave in a rational fashion. When they bite or otherwise offend, it is almost always because from their point of view it is the logical and right thing to do. The animal owner's attitude, if he is to succeed, must be one of patience, calm and willingness to understand. Discipline is necessary—not for the animal—but for his own human and thoroughly irrational temper.

Brief Bibliography

MAMMALS

Bourlière, François: *The Natural History of Mammals*. Alfred A. Knopf, New York, 1954.

Burt, W. H., and Grossenheider, R. P.: *A Field Guide to the Mammals*. Houghton Mifflin Co., Boston, 1952.

Calahane, Victor H.: *Mammals of North America*. The Macmillan Co., New York, 1954.

Dobie, J. Frank: *Voice of the Coyote*. University of Nebraska Press, Lincoln, 1961.

BIRDS

Hickey, Joseph J.: *A Guide to Bird Watching*. Oxford University Press, New York, 1943.

Peterson, Roger Tory: *A Field Guide to Birds* and *A Field Guide to Western Birds*. Houghton Mifflin Co., Boston, 1947 and 1941.

Wing, Leonard W.: *Natural History of Birds*. Ronald Press, New York, 1956.

REPTILES AND AMPHIBIANS

Conant, Roger: *A Field Guide to Reptiles and Amphibians*. Houghton Mifflin Co., Boston, 1958.

Noble, G. K.: *The Biology of the Amphibia*. Dover Publications, New York, 1954.

Oliver, James A.: *The Natural History of North American Amphibians and Reptiles*. D. Van Nostrand Co., Princeton, 1955.

Pope, Clifford H.: *The Reptile World*. Alfred A. Knopf, New York, 1955.

Schmidt, K. P., and Davis, D. Dwight, *Field Book of Snakes of the United States and Canada.* G. P. Putnam's Sons, New York, 1941.

Stebbins, Robert C.: *Amphibians and Reptiles of Western North America.* McGraw-Hill Book Co., New York, 1954.

ANIMAL BEHAVIOR

Fox, H. Munro: *The Personality of Animals.* New American Library, New York, 1947.

Hediger, H.: *Wild Animals in Captivity* and *Studies of the Psychology and Behaviour of Captive Animals in Zoos and Circuses.* Butterworth's Scientific Publications, London, 1950 and 1955.

Lorenz, Konrad Z.: *King Solomon's Ring.* Thomas Crowell Co., New York, 1952.

Scott, John Paul: *Animal Behavior.* University of Chicago Press, Chicago, 1958.

Tinbergen, N.: *Social Behaviour in Animals.* John Wiley and Sons, New York, 1953.

ECOLOGY AND MAN'S EFFECTS ON THE ENVIRONMENT

Carson, Rachel: *Silent Spring.* Houghton Mifflin Co., Boston, 1962.

Elton, Charles: *Animal Ecology.* Macmillan Co., New York, 1947.

Odum, Eugene P.: *Fundamentals of Ecology.* W. B. Saunders Co., Philadelphia, 1959.

Watts, May T.: *Reading the Landscape. An Adventure in Ecology.* Macmillan Co., New York, 1957.

Index

241

A CATALOGUE OF SELECTED DOVER BOOKS
IN ALL FIELDS OF INTEREST

A CATALOGUE OF SELECTED DOVER
BOOKS IN ALL FIELDS OF INTEREST

CELESTIAL OBJECTS FOR COMMON TELESCOPES, T. W. Webb. The most used book in amateur astronomy: inestimable aid for locating and identifying nearly 4,000 celestial objects. Edited, updated by Margaret W. Mayall. 77 illustrations. Total of 645pp. 5⅜ x 8½.
20917-2, 20918-0 Pa., Two-vol. set $8.00

HISTORICAL STUDIES IN THE LANGUAGE OF CHEMISTRY, M. P. Crosland. The important part language has played in the development of chemistry from the symbolism of alchemy to the adoption of systematic nomenclature in 1892. ". . . wholeheartedly recommended,"—Science. 15 illustrations. 416pp. of text. 5⅝ x 8¼.
63702-6 Pa. $6.00

BURNHAM'S CELESTIAL HANDBOOK, Robert Burnham, Jr. Thorough, readable guide to the stars beyond our solar system. Exhaustive treatment, fully illustrated. Breakdown is alphabetical by constellation: Andromeda to Cetus in Vol. 1; Chamaeleon to Orion in Vol. 2; and Pavo to Vulpecula in Vol. 3. Hundreds of illustrations. Total of about 2000pp. 6⅛ x 9¼.
23567-X, 23568-8, 23673-0 Pa., Three-vol. set $26.85

THEORY OF WING SECTIONS: INCLUDING A SUMMARY OF AIR-FOIL DATA, Ira H. Abbott and A. E. von Doenhoff. Concise compilation of subatomic aerodynamic characteristics of modern NASA wing sections, plus description of theory. 350pp. of tables. 693pp. 5⅜ x 8½.
60586-8 Pa. $6.50

DE RE METALLICA, Georgius Agricola. Translated by Herbert C. Hoover and Lou H. Hoover. The famous Hoover translation of greatest treatise on technological chemistry, engineering, geology, mining of early modern times (1556). All 289 original woodcuts. 638pp. 6¾ x 11.
60006-8 Clothbd. $17.50

THE ORIGIN OF CONTINENTS AND OCEANS, Alfred Wegener. One of the most influential, most controversial books in science, the classic statement for continental drift. Full 1966 translation of Wegener's final (1929) version. 64 illustrations. 246pp. 5⅜ x 8½.   61708-4 Pa. $3.00

THE PRINCIPLES OF PSYCHOLOGY, William James. Famous long course complete, unabridged. Stream of thought, time perception, memory, experimental methods; great work decades ahead of its time. Still valid, useful; read in many classes. 94 figures. Total of 1391pp. 5⅜ x 8½.
20381-6, 20382-4 Pa., Two-vol. set $13.00

THE DEPRESSION YEARS AS PHOTOGRAPHED BY ARTHUR ROTH-
STEIN, Arthur Rothstein. First collection devoted entirely to the work of
outstanding 1930s photographer: famous dust storm photo, ragged children,
unemployed, etc. 120 photographs. Captions. 119pp. 9¼ x 10¾.
23590-4 Pa. $5.00

CAMERA WORK: A PICTORIAL GUIDE, Alfred Stieglitz. All 559 illustrations and plates from the most important periodical in the history of
art photography, Camera Work (1903-17). Presented four to a page, reduced in size but still clear, in strict chronological order, with complete
captions. Three indexes. Glossary. Bibliography. 176pp. 8⅜ x 11¼.
23591-2 Pa. $6.95

ALVIN LANGDON COBURN, PHOTOGRAPHER, Alvin L. Coburn. Revealing autobiography by one of greatest photographers of 20th century
gives insider's version of Photo-Secession, plus comments on his own work.
77 photographs by Coburn. Edited by Helmut and Alison Gernsheim.
160pp. 8⅛ x 11.
23685-4 Pa. $6.00

NEW YORK IN THE FORTIES, Andreas Feininger. 162 brilliant photographs by the well-known photographer, formerly with Life magazine, show
commuters, shoppers, Times Square at night, Harlem nightclub, Lower
East Side, etc. Introduction and full captions by John von Hartz. 181pp.
9¼ x 10¾.
23585-8 Pa. $6.00

GREAT NEWS PHOTOS AND THE STORIES BEHIND THEM, John
Faber. Dramatic volume of 140 great news photos, 1855 through 1976,
and revealing stories behind them, with both historical and technical information. Hindenburg disaster, shooting of Oswald, nomination of Jimmy
Carter, etc. 160pp. 8¼ x 11.
23667-6 Pa. $5.00

THE ART OF THE CINEMATOGRAPHER, Leonard Maltin. Survey of
American cinematography history and anecdotal interviews with 5 masters—
Arthur Miller, Hal Mohr, Hal Rosson, Lucien Ballard, and Conrad Hall.
Very large selection of behind-the-scenes production photos. 105 photographs. Filmographies. Index. Originally Behind the Camera. 144pp.
8¼ x 11.
23686-2 Pa. $5.00

DESIGNS FOR THE THREE-CORNERED HAT (LE TRICORNE),
Pablo Picasso. 32 fabulously rare drawings—including 31 color illustrations
of costumes and accessories—for 1919 production of famous ballet. Edited
by Parmenia Migel, who has written new introduction. 48pp. 9⅜ x 12¼.
(Available in U.S. only)
23709-5 Pa. $5.00

NOTES OF A FILM DIRECTOR, Sergei Eisenstein. Greatest Russian
filmmaker explains montage, making of Alexander Nevsky, aesthetics; comments on self, associates, great rivals (Chaplin), similar material. 78 illustrations. 240pp. 5⅜ x 8½.
22392-2 Pa. $4.50

YUCATAN BEFORE AND AFTER THE CONQUEST, Diego de Landa. First English translation of basic book in Maya studies, the only significant account of Yucatan written in the early post-Conquest era. Translated by distinguished Maya scholar William Gates. Appendices, introduction, 4 maps and over 120 illustrations added by translator. 162pp. 5⅜ x 8½.
23622-6 Pa. $3.00

THE MALAY ARCHIPELAGO, Alfred R. Wallace. Spirited travel account by one of founders of modern biology. Touches on zoology, botany, ethnography, geography, and geology. 62 illustrations, maps. 515pp. 5⅜ x 8½.
20187-2 Pa. $6.95

THE DISCOVERY OF THE TOMB OF TUTANKHAMEN, Howard Carter, A. C. Mace. Accompany Carter in the thrill of discovery, as ruined passage suddenly reveals unique, untouched, fabulously rich tomb. Fascinating account, with 106 illustrations. New introduction by J. M. White. Total of 382pp. 5⅜ x 8½. (Available in U.S. only) 23500-9 Pa. $4.00

THE WORLD'S GREATEST SPEECHES, edited by Lewis Copeland and Lawrence W. Lamm. Vast collection of 278 speeches from Greeks up to present. Powerful and effective models; unique look at history. Revised to 1970. Indices. 842pp. 5⅜ x 8½. 20468-5 Pa. $6.95

THE 100 GREATEST ADVERTISEMENTS, Julian Watkins. The priceless ingredient; His master's voice; 99 44/100% pure; over 100 others. How they were written, their impact, etc. Remarkable record. 130 illustrations. 233pp. 7⅞ x 10 3/5. 20540-1 Pa. $5.00

CRUICKSHANK PRINTS FOR HAND COLORING, George Cruickshank. 18 illustrations, one side of a page, on fine-quality paper suitable for watercolors. Caricatures of people in society (c. 1820) full of trenchant wit. Very large format. 32pp. 11 x 16. 23684-6 Pa. $4.50

THIRTY-TWO COLOR POSTCARDS OF TWENTIETH-CENTURY AMERICAN ART, Whitney Museum of American Art. Reproduced in full color in postcard form are 31 art works and one shot of the museum. Calder, Hopper, Rauschenberg, others. Detachable. 16pp. 8¼ x 11.
23629-3 Pa. $2.50

MUSIC OF THE SPHERES: THE MATERIAL UNIVERSE FROM ATOM TO QUASAR SIMPLY EXPLAINED, Guy Murchie. Planets, stars, geology, atoms, radiation, relativity, quantum theory, light, antimatter, similar topics. 319 figures. 664pp. 5⅜ x 8½.
21809-0, 21810-4 Pa., Two-vol. set $10.00

EINSTEIN'S THEORY OF RELATIVITY, Max Born. Finest semi-technical account; covers Einstein, Lorentz, Minkowski, and others, with much detail, much explanation of ideas and math not readily available elsewhere on this level. For student, non-specialist. 376pp. 5⅜ x 8½.
60769-0 Pa. $4.00

AMERICAN BIRD ENGRAVINGS, Alexander Wilson et al. All 76 plates. from Wilson's *American Ornithology* (1808-14), most important orthnitho-logical work before Audubon, plus 27 plates from the supplement (1825-33) by Charles Bonaparte. Over 250 birds portrayed. 8 plates also reproduced in full color. 111pp. 9⅜ x 12½. 23195-X Pa. $6.00

CRUICKSHANK'S PHOTOGRAPHS OF BIRDS OF AMERICA, Allan D. Cruickshank. Great ornithologist, photographer presents 177 closeups, groupings, panoramas, flightings, etc., of about 150 different birds. Ex-panded *Wings in the Wilderness*. Introduction by Helen G. Cruickshank. 191pp. 8¼ x 11. 23497-5 Pa. $6.00

AMERICAN WILDLIFE AND PLANTS, A. C. Martin, et al. Describes food habits of more than 1000 species of mammals, birds, fish. Special treatment of important food plants. Over 300 illustrations. 500pp. 5⅜ x 8½. 20793-5 Pa. $4.95

THE PEOPLE CALLED SHAKERS, Edward D. Andrews. Lifetime of research, definitive study of Shakers: origins, beliefs, practices, dances, social organization, furniture and crafts, impact on 19th-century USA, present heritage. Indispensable to student of American history, collector. 33 illustrations. 351pp. 5⅜ x 8½. 21081-2 Pa. $4.00

OLD NEW YORK IN EARLY PHOTOGRAPHS, Mary Black. New York City as it was in 1853-1901, through 196 wonderful photographs from N.-Y. Historical Society. Great Blizzard, Lincoln's funeral procession, great buildings. 228pp. 9 x 12. 22907-6 Pa. $7.95

MR. LINCOLN'S CAMERA MAN: MATHEW BRADY, Roy Meredith. Over 300 Brady photos reproduced directly from original negatives, photos. Jackson, Webster, Grant, Lee, Carnegie, Barnum; Lincoln; Battle Smoke, Death of Rebel Sniper, Atlanta Just After Capture. Lively com-mentary. 368pp. 8⅜ x 11¼. 23021-X Pa. $6.95

TRAVELS OF WILLIAM BARTRAM, William Bartram. From 1773-8, Bartram explored Northern Florida, Georgia, Carolinas, and reported on wild life, plants, Indians, early settlers. Basic account for period, enter-taining reading. Edited by Mark Van Doren. 13 illustrations. 141pp. 5⅜ x 8½. 20013-2 Pa. $4.50

THE GENTLEMAN AND CABINET MAKER'S DIRECTOR, Thomas Chippendale. Full reprint, 1762 style book, most influential of all time; chairs, tables, sofas, mirrors, cabinets, etc. 200 plates, plus 24 photographs of surviving pieces. 249pp. 9⅞ x 12¾. 21601-2 Pa. $6.50

AMERICAN CARRIAGES, SLEIGHS, SULKIES AND CARTS, edited by Don H. Berkebile. 168 Victorian illustrations from catalogues, trade journals, fully captioned. Useful for artists. Author is Assoc. Curator, Div. of Trans-portation of Smithsonian Institution. 168pp. 8½ x 9½. 23328-6 Pa. $5.00

MUSHROOMS, EDIBLE AND OTHERWISE, Miron E. Hard. Profusely illustrated, very useful guide to over 500 species of mushrooms growing in the Midwest and East. Nomenclature updated to 1976. 505 illustrations. 628pp. 6½ x 9¼. 23309-X Pa. $7.95

AN ILLUSTRATED FLORA OF THE NORTHERN UNITED STATES AND CANADA, Nathaniel L. Britton, Addison Brown. Encyclopedic work covers 4666 species, ferns on up. Everything. Full botanical information, illustration for each. This earlier edition is preferred by many to more recent revisions. 1913 edition. Over 4000 illustrations, total of 2087pp. 6⅛ x 9¼. 22642-5, 22643-3, 22644-1 Pa., Three-vol. set $24.00

MANUAL OF THE GRASSES OF THE UNITED STATES, A. S. Hitchcock, U.S. Dept. of Agriculture. The basic study of American grasses, both indigenous and escapes, cultivated and wild. Over 1400 species. Full descriptions, information. Over 1100 maps, illustrations. Total of 1051pp. 5⅜ x 8½. 22717-0, 22718-9 Pa., Two-vol. set $12.00

THE CACTACEAE,, Nathaniel L. Britton, John N. Rose. Exhaustive, definitive. Every cactus in the world. Full botanical descriptions. Thorough statement of nomenclatures, habitat, detailed finding keys. The one book needed by every cactus enthusiast. Over 1275 illustrations. Total of 1080pp. 8 x 10¼. 21191-6, 21192-4 Clothbd., Two-vol. set $35.00

AMERICAN MEDICINAL PLANTS, Charles F. Millspaugh. Full descriptions, 180 plants covered: history; physical description; methods of preparation with all chemical constituents extracted; all claimed curative or adverse effects. 180 full-page plates. Classification table. 804pp. 6½ x 9¼. 23034-1 Pa. $10.00

A MODERN HERBAL, Margaret Grieve. Much the fullest, most exact, most useful compilation of herbal material. Gigantic alphabetical encyclopedia, from aconite to zedoary, gives botanical information, medical properties, folklore, economic uses, and much else. Indispensable to serious reader. 161 illustrations. 888pp. 6½ x 9¼. (Available in U.S. only) 22798-7, 22799-5 Pa., Two-vol. set $11.00

THE HERBAL or GENERAL HISTORY OF PLANTS, John Gerard. The 1633 edition revised and enlarged by Thomas Johnson. Containing almost 2850 plant descriptions and 2705 superb illustrations, Gerard's *Herbal* is a monumental work, the book all modern English herbals are derived from, the one herbal every serious enthusiast should have in its entirety. Original editions are worth perhaps $750. 1678pp. 8½ x 12¼. 23147-X Clothbd. $50.00

MANUAL OF THE TREES OF NORTH AMERICA, Charles S. Sargent. The basic survey of every native tree and tree-like shrub, 717 species in all. Extremely full descriptions, information on habitat, growth, locales, economics, etc. Necessary to every serious tree lover. Over 100 finding keys. 783 illustrations. Total of 986pp. 5⅜ x 8½. 20277-1, 20278-X Pa., Two-vol. set $10.00

GEOMETRY, RELATIVITY AND THE FOURTH DIMENSION, Rudolf Rucker. Exposition of fourth dimension, means of visualization, concepts of relativity as Flatland characters continue adventures. Popular, easily followed yet accurate, profound. 141 illustrations. 133pp. 5⅜ x 8½.
23400-2 Pa. $2.75

THE ORIGIN OF LIFE, A. I. Oparin. Modern classic in biochemistry, the first rigorous examination of possible evolution of life from nitrocarbon compounds. Non-technical, easily followed. Total of 295pp. 5⅜ x 8½.
60213-3 Pa. $4.00

THE CURVES OF LIFE, Theodore A. Cook. Examination of shells, leaves, horns, human body, art, etc., in *"the* classic reference on how the golden ratio applies to spirals and helices in nature . . . ."—Martin Gardner. 426 illustrations. Total of 512pp. 5⅜ x 8½. 23701-X Pa. $5.95

PLANETS, STARS AND GALAXIES, A. E. Fanning. Comprehensive introductory survey: the sun, solar system, stars, galaxies, universe, cosmology; quasars, radio stars, etc. 24pp. of photographs. 189pp. 5⅜ x 8½. (Available in U.S. only) 21680-2 Pa. $3.00

THE THIRTEEN BOOKS OF EUCLID'S ELEMENTS, translated with introduction and commentary by Sir Thomas L. Heath. Definitive edition. Textual and linguistic notes, mathematical analysis, 2500 years of critical commentary. Do not confuse with abridged school editions. Total of 1414pp. 5⅜ x 8½. 60088-2, 60089-0, 60090-4 Pa., Three-vol. set $18.00

DIALOGUES CONCERNING TWO NEW SCIENCES, Galileo Galilei. Encompassing 30 years of experiment and thought, these dialogues deal with geometric demonstrations of fracture of solid bodies, cohesion, leverage, speed of light and sound, pendulums, falling bodies, accelerated motion, etc. 300pp. 5⅜ x 8½. 60099-8 Pa. $4.00